T0155428

On Literature and Science

On Literature and Science

Essays, Reflections, Provocations

Philip Coleman
EDITOR

FOUR COURTS PRESS

Set in 10.5 on 13 point AGaramond for
FOUR COURTS PRESS LTD
7 Malpas Street, Dublin 8, Ireland
e-mail: info@fourcourtspress.ie
http://www.fourcourtspress.ie
and in North America for
FOUR COURTS PRESS
c/o ISBS, 920 N.E. 58th Avenue, Suite 300, Portland, OR 97213.

ISBN 978-1-84682-071-7

A catalogue record for this title
is available from the British Library.

Printed in England
by Antony Rowe Ltd, Chippenham, Wilts.

for the Traas family, Tipperary

'en langzaam begin ik te voelen en te denken
dat ook de boomgard daarnaar zoekt – dat wij
hetzelfde zoeken, de boomgaard en ek'

Rutger Kopland

Contents

Foreword

JOHN HEGARTY (PROVOST, TRINITY COLLEGE DUBLIN)

Was Leonardo da Vinci a scientist or an artist? Is the question even relevant? His was a mind that encompassed all knowledge in a seamless way. Today, when the 'arts and humanities' and the 'sciences' are addressed in the same sentence, it is often in terms of the tension and competition between them. Public debate, insofar as it addresses the breadth of human knowledge and endeavour, tends to be polarized, either stressing the importance of science only to the economy, or bemoaning the downgrading of the arts and humanities with respect to the sciences.

While this discord may make for good after-dinner debate, it misses the point. The sciences and the arts/humanities are equally about creativity and the imagination. All of these disciplines are essential to our understanding of the society in which we live and its overall wellbeing. Happily for us, modern life allows for unprecedented interplay between these disciplines, leading to the possibility of ever more exciting outcomes.

In 2005, Trinity College and the city of Dublin hosted the British Association Festival of Science. For a week both city and campus were abuzz with the unbridled enthusiasm of children and adults of all ages and backgrounds, as they embraced the opportunity to learn more about the latest developments in science. Equally impressive was that for the duration of the Festival the TCD School of English organized a series of exciting lectures on the connections between literature and science. To me, that represents one of the unique characteristics of the university – the opportunity that exists to creatively explore ideas that span disciplines and domains of knowledge.

The essays in this new book are reflections on science and on the connections between science and literature from a variety of authors. The essays trace how writers in the past grappled with the concepts of science and the effect of scientific developments on the course of human civilization. Some address science as it is evolving today while others pose – and answer – the frequently asked question of whether literature and science are true bedfellows.

What emerges with great clarity is the recognition by the authors that, from an intellectual and emotional point of view, each individual needs to feel comfortable with the totality of their experience. Universities work on the basis of disciplines, but it is the synthesis of the knowledge arising from all disciplines that society seeks now more than ever. This book addresses that vital human need and its message will surely find a great resonance in the minds of those who read it.

Acknowledgments

This book gathers together contributions to two series of public lectures and talks held in Trinity College Dublin in 2005. I would like to thank my colleagues in the School of English for supporting the lecture series on literature and science I organised in Michaelmas term of that year, and the British Association for the Advancement of Science for including a series of lunchtime talks in its 'Science in the City' festival programme that summer. Lilian Foley, Stephen Matterson, and Diane Sadler, provided assistance on both occasions, for which I am very grateful. Philippa Harris, Festival of Science officer, was extremely helpful in coordinating the lunchtime lecture series, as was Joe Carroll, former Dean of Science in TCD.

I wish to thank Michael Adams of Four Courts Press for agreeing to publish this book, Martin Fanning for his invaluable editorial assistance, and Anthony Tierney and Aoife Walsh for their help in seeing this book through its various stages of production. I thank all of the contributors for their patience while the book has been prepared for publication, and especially those who came on board in the later stages of its development. I regret that Nicholas Daly and Aileen Douglas were unable to contribute to this volume: their talks on Robert Louis Stevenson and Jonathan Swift, respectively, were among the highlights of the first literature and science lecture series. Many members of the College community attended all of these lectures, and some of them have contributed to this volume. I am grateful, in particular, to Iggy McGovern, John Scattergood, George Sevastopulo, and Ross Skelton for their support and encouragement. Harry Clifton, Kit Fryatt, and Harry Gilonis provided much-appreciated advice which I am also happy to acknowledge here. I am extremely grateful to Oliver Jeffers for granting permission to reproduce his painting 'Portrait with Russell and Whitehead (Worlds Apart?)' on the cover, and to the School of English and the School of Physics, Trinity College Dublin, for their financial assistance in this regard. This book would not have been published without the support of John Hegarty, Provost of Trinity College Dublin. In an article published in the *Irish Times* in November 2006, Dr Hegarty wrote: 'A great university is one in which there is a balance between the humanities and the sciences across a wide sweep of disciplines and across the breadth of human knowledge and experience.' His support for this project – as a member of the audience on a number of occasions, as the author of the Foreword to this volume, and for providing financial assistance in the form of an award from the Provost's Academic Development Fund – has been crucial, and it illustrates his belief in the value of research in the arts and humanities not just in Trinity but in the modern university.

On Literature and Science is dedicated to the Traas family, Cahir, Co. Tipperary, and in particular to Willem, Alie, and Con, without whose friendship and support I could not have pursued my undergraduate and early postgraduate studies in English literature in the 1990s.

Introduction

PHILIP COLEMAN

Great is language … it is the mightiest of the sciences.

Walt Whitman[1]

This book has its basis in two separate but related events held in Trinity College Dublin in 2005. In Michaelmas term of that year the School of English organized a public lecture series on literature and science, with contributions from Nicholas Daly, Aileen Douglas, Kate Hebblethwaite, Darryl Jones, Iggy McGovern, Stephen Matterson, Amanda Piesse, John Scattergood, and myself. Following on the success of that series I organized five lunchtime lectures and readings as part of the British Association for the Advancement of Science festival 'Science in the City' which was held in Dublin that summer and included an inspirational contribution from the poet Harry Clifton. In an important sense, however, this book gathers together the thoughts and reflections of more than those whose work is represented between its covers, some of whom did not contribute in a direct way to either of those initial events. Helen Conrad-O'Briain, Allen Fisher, Kit Fryatt, Dylan Harris, Randolph Healy, Benjamin Keatinge, Peter Middleton, Andrew J. Power, Maurice Scully, Ross Skelton, and Meredith Quartermain did not give lectures or readings on these occasions, but in one way or another they were there from the beginning – in the audience, corresponding by email as the book developed, sharing ideas and writings that broaden our sense of the relationship between literature and science. Many of the points raised by members of the public and the broader academic community in College during those original events have also been taken on board and are now incorporated in the versions of those lectures and readings printed here. It is true to say, then, that this book is both a permanent record and a continuation of the lively and sometimes intense conversations that were stimulated by many of the talks on literature and science held in TCD in 2005, and I believe it will provide a stimulus for further debate in the future.

1 Walt Whitman, *Leaves of Grass & other writings*, ed. Michael Moon (New York and London, 2002), p. 750.

Each series of lectures and talks was organized with one basic aim in mind: to explore some of the ways that science and scientific ideas have been engaged by writers of prose, poetry, and drama, from medieval times to the present. The issue of whether there is any common ground between the two areas has often been the subject of controversy and intractable disagreement. In recent times it has been most famously summarized by C.P. Snow's description of 'the two cultures' in a 1956 essay later elaborated in his Rede Lecture at the University of Cambridge in 1959. Snow identified the 'problem' in the following terms:

> I believe the intellectual life of the whole of western society is increasingly being split into two polar groups. When I say the intellectual life, I mean to include also a large part of our practical life, because I should be the last person to suggest the two can at the deepest level be distinguished. [...] Two polar groups: [...] Literary intellectuals at one pole – at the other scientists, and as the most representative, the physical scientists.[2]

Snow's lecture explored some of the ways that British universities in the 1960s could narrow the gap between these 'two polar groups' but, as he conceded in a postscript written in 1963: 'Changes in education are not going to produce miracles.'[3] Nevertheless, he argued that what he called 'communications' between the two groups could be improved, with the ultimate goal of educating 'a large proportion of our better minds so that they are not so ignorant of imaginative experience, both in the arts and in science, nor ignorant either of the endowments of applied science'.[4] Snow's words call to mind the Scottish poet Hugh MacDiarmid's poem 'Poetry and Science', in which he argues that 'Nature is more wonderful / When it is at least partly understood.' 'The rarity and value of scientific knowledge / Is little understood' his poem begins, and continues:

> – even as people
> Who are not botanists find it hard to believe
> Special knowledge of the subject can add
> Enormously to the aesthetic appreciation of flowers![5]

For MacDiarmid, as for Snow, it was crucial that people should be aware of many different ways of apprehending the world – scientific and aesthetic – and the key

2 C.P. Snow, *The two cultures: and a second look; an expanded version of the two cultures and the scientific revolution* (Cambridge, 1964), pp 3–4. In the 1940s the great scholar of W.B. Yeats' poetry, T.R. Henn, gave a series of lectures on aspects of English poetry designed specifically for science students at Cambridge University. His lectures are collected in *The apple and the spectroscope; being lectures on poetry designed (in the main) for science students* (London, 1951). Today the Broad Curriculum initiative in Trinity College Dublin might be said to fulfil a similar function by providing introductory courses in certain areas of the arts and humanities (such as English literature, philosophy, film studies, and the history of art) for students in the sciences. 3 Snow, p. 100. 4 Ibid. 5 Hugh MacDiarmid, *Selected poems* (London, 1992), p. 228.

to increased understanding in both areas was a matter of improving the channels of communication between them.

In recent years, certainly, there has been a great deal of willingness to explore the connections between literature and science, despite the fact that many influential twentieth-century thinkers, including the British physicist Paul Dirac and the French philosopher Gaston Bachelard, asserted the kind of polar oppositeness disputed by Snow. Dirac famously suggested that physics and poetry are 'in opposition',[6] while Bachelard, in his book *The psychoanalysis of fire* (1938), argued that:

> The axes of poetry and of science are opposed to one another from the outset. All that philosophy can hope to accomplish is to make poetry and science complementary, to unite them as two well-defined opposites. We must oppose then, to the enthusiastic, poetic mind the taciturn, scientific mind, and for the scientific mind an attitude of preliminary antipathy is a healthy precaution.[7]

As intriguing and innovative as many of Bachelard's ideas were and indeed still are today, his claim about the gap between poetry and science has been challenged by many more recent thinkers. More importantly, the categories delineated here have been challenged by women and men who have demonstrated an ability not only to merge the two opposing axes of thought described by Dirac and Bachelard, but to render them almost indistinguishable from each other – from pure scientists and mathematicians such as Graham Farmelo and Steven Weinberg who strive after 'beautiful' equations,[8] to poets such as Mei-mei Berssenbrugge and Elizabeth Willis, who have produced stunning bodies of work that demonstrate a keen awareness of scientific *and* aesthetic procedures and ideas.[9] In fact, it could be said that Bachelard's statement now represents an increasingly conservative view, with more and more intellectuals and teachers of literature and science today positing the importance of broad awareness in the arts and sciences in addi-

6 Quoted at http://www.dirac.ch/PaulDirac.html, accessed 10 May 2007. Dirac apparently made this claim when he heard that fellow-physicist J. Robert Oppenheimer had started writing poetry. **7** Gaston Bachelard, *The psychoanalysis of fire*, trans. Alan C.M. Ross (London and New York, 1987), p. 2. **8** See Graham Farmelo (ed.), *It must be beautiful: great equations of modern science* (London, 2002). In an online poll conducted in 2004, the equation voted the most beautiful was Euler's identity equation: $e^{i\pi/2} = i$. As one participant explained: 'The equation contains nine basic concepts of mathematics – once and only once – in a single expression. These are: *e* (the base of natural logarithms); the exponent operation; pi; plus (or minus, depending on how you write it); multiplication; imaginary numbers; equals; one; and zero.' See http://www.collisiondetection.net/mt/archives/001013.html, accessed 10 May 2007. **9** See, for example, Mei-mei Berssenbrugge, *Endocrinology* (Berkeley, CA, 1997) and Elizabeth Willis, *Meteoric Flowers* (Middletown, CT, 2006). Willis's poetry engages with the work of Erasmus Darwin (1731–1802), eighteenth-century scientist and grandfather of Charles Darwin, who attempted to describe his findings in zoology, botany, and other fields in a number of long poetic works. These include *The Botanic Garden* (London, 1791), in which he claims the 'general design is to enlist Imagination under the banner of Science' (p. v), and *Temple of Nature* (London, 1803).

tion to one's own scholarly specialism. No-one can be a specialist in everything, but the radical separation of different fields of inquiry suggested by Bachelard and others may eventually be seen as a thing of the past.

It needs to be acknowledged, of course, that artists and scientists have long challenged the kinds of intellectual bifurcation suggested by Snow's idea of 'the two cultures' and Bachelard's apparently irreconcilable 'axes of poetry and science.' As the essays in the first part of this book clearly demonstrate, contact and collaboration between what we now call the arts and sciences go back long before the medieval period, and writers have always sought to engage new ideas – scientific or otherwise – in their work, from Geoffrey Chaucer to George Saunders. What is at stake here, then, is not so much the issue of whether writers ever write about science, but if it is possible to say that writers and scientists can share the same view of the world. The chemist Peter Atkins has argued that:

> Although poets may aspire to understanding, their talents are more akin to entertaining self-deception. They may be able to emphasise delights in the world, but they are deluded if they and their admirers believe that their identification of delights and their use of poignant language are enough for comprehension. Philosophers too, I am afraid, have contributed to the understanding of the universe little more than poets. [...] They have not contributed much that is novel until after novelty has been discovered by scientists. [...] While poetry titillates and theology obfuscates, science liberates.[10]

Atkins clearly would not have agreed with T.S. Eliot's claim that the publication of James Joyce's *Ulysses* in 1922 had 'the importance of a scientific discovery'.[11] The philosopher Mary Midgley, however, has taken Atkins seriously and in her book *Science and poetry* she argues that he fails to recognize the common ground that may be said to exist between poetic (artistic/literary) and scientific forms of inquiry. More importantly, Midgley argues that the idea of 'knowledge' that informs Atkins' thinking is based on an acceptance of age-old binaries – 'Science v. Religion, Reason v. Feeling, Classical v. Romantic, Masculine v. Feminine, Materialist v. Idealist, Body v. Mind' – all of which 'share one unfortunate feature: they all tend to drive us into inner conflict.'[12] Against the view that one should always come down on one side at the expense of another, as Atkins does, Midgley suggests that we should search for a deeper 'unity' between different disciplines and resist the kind of 'fragmentation' of thought that ultimately leads to a breakdown of 'the unity between ourselves and the living world around us.'[13]

10 Peter Atkins, 'The limitless power of science' in John Cornwell (ed.), *Nature's imagination* (Oxford, 1995), p. 123. **11** T.S. Eliot, 'Ulysses, order, and myth' in Frank Kermode (ed.), *Selected prose of T.S. Eliot* (London, 1975), p. 177. **12** Mary Midgley, *Science and poetry* (2001), p. xi. **13** Ibid.

In a recently published collection of essays on contemporary poetry's engagements with contemporary science, Robert Crawford has dismissed Midgley's views. He suggests that poetry, for her, 'denote[s] those elements of life which we assume to be unscientific' and he admonishes her 'misty use of the word [poetry].'[14] The word 'poetry', however, like the word 'science', is semantically and culturally elusive.[15] Historical variations in meaning aside, Midgley believes one of the reasons for the confusion that surrounds these terms is the intellectual and academic conflicts that have always endeavoured to keep them apart. Against that view she advocates a form of understanding that acknowledges their mutual capacities for advancing self-understanding, a view that Crawford's book actually reinforces in its presentation of poets and scientists attempting to describe the same worldly phenomena. In a lecture delivered before the British Association for the Advancement of Science in Belfast in 1874, John Tyndall sought to describe the importance of literary skill in the propagation of scientific ideas. 'It has been said by its opponents that science divorces itself from literature', he argued, 'but the statement, like so many others arises from lack of knowledge':

> A glance at the less technical writings of its leaders [...] would show what breadth of literary culture they command. Where among modern writers can you find their superiors in clearness and vigour of literary style? Science desires not isolation, but freely combines with every effort towards the bettering of man's estate. Single-handed, and supported, not by outward sympathy, but by inward force, it has built at least one great wing of the many-mansioned home which man in his totality demands.[16]

Midgley also urges a consideration of the 'totality' of thought produced by literature and science signalled here by Tyndall, as did the English poet and critic William Empson. Empson, who began his university career as a student of mathematics before turning to English literature, suggested that:

> one can define terms as one likes, and it is possible to define meaning, significance, and thought as all occurring when there is a reference to human wishes or ideals as well as to facts. [...] However even accepting the definitions I should deny that they make science meaningless or insignificant or unthinking. One of our major desires about the universe is that it should be orderly and capable of being understood; this is partly for our

14 Robert Crawford (ed.), *Contemporary poetry and contemporary science* (Oxford, 2006), p. 3. **15** In the second edition of the *Oxford English Dictionary* (1989), the word 'poetry' is defined in six separate entries, 'science' in seven, and 'scientific' in six. 'Literature' has three separate entries. **16** John Tyndall, from 'The Belfast Address' (1874) in Laura Otis (ed.), *Literature and science in the nineteenth century: an anthology* (Oxford, 2002), p. 4.

> own safety and partly from a metaphysical feeling that we want to be at home in it or not alien from it. [...] Copernicus [...] has been accused of believing his theory because of its great elegance and simplicity in spite of the facts appearing to be against it; not because he was indifferent to truth but because he trusted his wish that the true answer should be elegant and simple. I do not mean that this is the whole story; we want a good deal more of the universe besides intelligibility, and a certain amount more I think is provided by science though not much; and on the other hand the desire to recognise the truth however disgusting it may be is itself an 'ideal'. So the dichotomy, it seems to me, breaks down both ways round; and for that matter the 'social consequences' of it are bad for science as well as for the humanities.[17]

Empson's suggestion that 'we want to be at home' in the universe, and that literature and science can both assuage this kind of existential longing, echoes Tyndall's depiction of a 'many-mansioned home which man in his totality demands.' Both men believed, as does Midgley, that between them the efforts of the scientist and the artist can lead to a fuller understanding of the universe, as well as a reconciliation between the 'two cultures' identified in the 1960s by C.P. Snow.

The positions held by Tyndall, Empson, and Midgley on the relationship between literature and science are summarized by the speaker of a poem called 'Ballad' by the American poet A.R. Ammons, who says: 'I want to know the unity in all things and the difference / between one thing and another'.[18] Ultimately, however, it could be said that language is the space wherein both the scientific and the poetic (or literary) reside – to take Tyndall's metaphor a step further. Long before the German philosopher Martin Heidegger wrote that 'Language is the house of Being' and the American critic Fredric Jameson posited the idea of 'the prison-house of language',[19] Walt Whitman, the grand modern poet of intellectual and imaginative expansiveness wrote:

> Great is language ... it is the mightiest of the sciences,
> It is the fullness and color and form and diversity of the earth ... and of men and women ... and of all qualities and processes;
> It is greater than wealth ... it is greater than building or ships or religions or paintings or music.[20]

17 William Empson, letter to Qien Xuexi, 17 September 1947, in John Haffenden (ed.), *Selected letters of William Empson* (Oxford, 2006), p. 152. **18** A.R. Ammons, *Selected poems* (New York, 2006), p. 62. **19** See Martin Heidegger, 'Letter on humanism' in David Farrell Krell (ed.), *Basic writings* (1977; rev. ed. London, 1993), p. 217, and Fredric Jameson, *The prison-house of language: a critical account of structuralism and Russian formalism* (Princeton, NJ, 1974). **20** Whitman, *Leaves of Grass & other writings*, p. 750.

In the prose preface to the first edition of *Leaves of Grass* (1855), Whitman argued that: 'Exact science and its practical movements are no checks on the greatest poet but always his encouragement and support.'[21] Throughout his work Whitman plays with the possibilities for understanding self and world that are presented by different kinds of knowledge, but he is the first to acknowledge that there are times when one form seems more appropriate or useful than another. In a poem entitled 'When I heard the learn'd astronomer', included in later editions of *Leaves of Grass,* the speaker turns his back on 'the proofs, the figures [...] ranged in columns before [him]' at a public lecture on astronomy, and he 'wander[s] off by [himself], / In the mystical moist night-air, and from time to time, / Look[s] up in perfect silence at the stars'.[22] While he grows 'tired and sick' of listening to one man lecturing on astronomy here, however, Whitman's speaker does not seek to deny the place of the scientific in explanations of the universe. Rather, he suggests that science – like literature – can sometimes send its audience back into a consideration of the natural world and those things with which both the writer and the scientist are fundamentally concerned. Indeed, Whitman argues that this is the end of all great thought – to bring us back into a closer relation with our origins: 'What has ever happened ... what happens and whatever may or shall happen, the vital law encloses all ...'[23]

II

This book, then, contains a number of contributions that may be said to support Mary Midgley's view of literature and science as forms of understanding that contribute in equally valuable ways to our sense of self and world. What distinguishes it from recent collections about the relationship between literature and science, however, is the range of topics and materials covered, as well as the historical scope of its contents. It is divided into two parts. The first contains eleven essays that explore literary engagements with science from the medieval period to the present, beginning with Helen Conrad-O'Briain's analysis of the influence of science on the work of Geoffrey Chaucer and ending with Peter Middleton's reflections on the relationships between modern poetry and scientific method. The second part of the book contains eight pieces – prose texts interfused with poetry by Iggy McGovern and Randolph Healy, and poems or sections from longer poetic texts by Meredith Quartermain, Harry Clifton, Allen Fisher, Maurice Scully, Dylan Harris, and Kit Fryatt. The inclusion of different kinds of writing in the book is intended to reflect the diversity of literary engagements with sci-

21 Ibid., p. 626. **22** Ibid., p. 227. The Canadian astronomer and poet Rebecca Elson, who died in 1999 at the age of 39, wrote a number of poems that explore this Whitmanian theme. See, for example, 'Girl with a Balloon' and 'Explaining Relativity' in *A Responsibility to Awe* (Manchester, 2001), pp 12, 13. **23** Whitman, *Leaves of Grass & other writings*, p. 626.

ence and the extent to which both scholars of literature and practitioners of various literary arts are engaged by scientific ideas.

Science fiction is perhaps the most well-known and influential form of literature in which writers have engaged with scientific ideas, and especially in the twentieth century. In her essay, however, Helen Conrad-O'Briain traces the origins of the genre back to the medieval period, arguing that Chaucer's 'Franklin's Tale' may be considered one of the earliest examples of a kind of writing about science that is often seen as a distinctly modern phenomenon. Conrad-O'Briain describes Chaucer's engagements with various technologies and scientific ideas – from the Ptolemaic view of the world to optics theory – and her essay provokes a radical reconsideration of the ways that science is read in relation to literary texts and representations. John Scattergood's essay continues this exploration of medieval literature by examining the different ways that writers from Dante to John Donne have incorporated ideas derived from new technologies in their work. Like Conrad-O'Briain, he stresses the close relationship between developments in science and the evolution of literary styles, forms, and subjects, and he illustrates this point with a brilliant close reading of Donne's elegy for John Harrington, 2nd Baron of Exton, written in 1614. The proximity between scientific and literary worldviews demonstrated by the book's opening essays is taken a step further in Amanda Piesse's detailed account of the relationship between the science of anatomy and literary practice during the Renaissance period. The title of her essay – 'Bodies of knowledge and knowledge of the body in sixteenth-century literature' – indicates the closeness of this relationship, but it also suggests that the origins of 'modern' thought were not as divided as they may seem today. Piesse argues that 'the scientific investigation of the body and the more metaphysical investigation of what it is to be "human" never really separates out fully during the sixteenth century, and this is as true of scientific works as it is of literary works.' In making this claim and examining it in relation to a wide range of literary and scientific texts, she introduces one of the book's major themes – the relationship between literature and medical science – which is treated in different ways by Andrew J. Power, Benjamin Keatinge, and Ross Skelton in their contributions to the collection.

Described by the Polish poet and Nobel Laureate Czesław Miłosz as 'a doctor with profound knowledge of the humanities, in whom poets and artists sense a kindred mentality,'[24] the medical scientist Andrzej Szczeklik has argued that 'Medicine and art are descended from the same roots. They both originated in magic – a practice based on the omnipotence of the word.'[25] In his essay here, Andrew J. Power explores the character of Hamlet in terms of sixteenth- and sev-

24 Czesław Miłosz, 'Foreword' to Andrzej Szczeklik, *Catharsis: on the art of medicine*, trans. Antonia Lloyd-Jones (Chicago and London, 2005), p. ix. **25** Szczeklik, p. 9.

enteenth-century medical theory. The magic of Shakespeare's language in *Hamlet* and other plays reveals a close engagement with Renaissance science, he argues, which serves to highlight disorder within the body of the subject (Prince Hamlet) and the broader body politic. This relationship between private and public health, as it were, and the impact of scientific ideas on literary or artistic representations of both, is explored in more recent contexts by Benjamin Keatinge and Ross Skelton in essays that examine the various ways that psychoanalysis – a relatively recent and controversial addition to medical science – has been engaged by writers and painters from a variety of cultural contexts. In an essay that describes his own discovery of the importance of the area of psychoanalysis known as 'bilogic', Skelton focuses on deep strands and patterns in the work of Irish poet Louis MacNeice, while Keatinge takes a broader view of the relationship between early developments in psychoanalysis and the aesthetic and literary impact they had on surrealism. Exploring the work of a wide range of visual and literary artists, from Francis Picabia to André Breton, he shows how developments in one field of (medical) science can impact on cultural production, a point that is explored with regard to many different kinds of scientific inquiry in the essays gathered here by Kate Hebblethwaite, Darryl Jones, and Stephen Matterson.

In her essay, Kate Hebblethwaite provides a detailed description of the kinds of scientific debates that informed nineteenth-century (popular) literary culture, and she shows how they may be traced across a range of cultural productions, from *Punch* cartoons to the publication of Charles Kingsley's *The Water Babies* in 1863 and beyond. Focussing on visual and textual representations of the hippopotamus, she argues that 'the fulsome figure of the hippo served a hitherto-unacknowledged function as an important gauge for shifting attitudes about evolution theory', and thereby provides an illuminating insight into the efficacy of the (literary) image in the propagation of scientific ideas. The imagery of science is one of the topics explored by Darryl Jones in his essay on the 'imagination of disaster' and mass-death in modern literature, a body of writing that has frequently been influenced by developments in scientific thought. In an essay that includes descriptions of contemporary films and writings that engage with these ideas, Jones focuses on the work of H.G. Wells, a writer of science fiction whose vision of the world, he argues, is less fantastic than is often suggested. Edgar Allan Poe's work has often been discredited for its apparent retreat into the realm of fantasy, but in his essay Stephen Matterson shows how closely Poe's fictions engage with nineteenth-century developments in science, suggesting that both the artist and the scientist 'must see beneath the offered surface to find truth, [to] demonstrate the capacity to re-imagine reality.' This is a point that I explore also in my contribution to the collection, in relation to a wide range of American short story writers from the nineteenth century to the present, including Poe, Nathaniel Hawthorne, David Foster Wallace, and George Saunders.

Towards the end of Henry James's tale 'Daisy Miller' the character of Winterbourne encounters Daisy and her Italian 'guide' Giovanelli in the Colosseum in Rome. 'The place had never seemed to him more impressive', we are told, and as Winterbourne 'stood there he began to murmur Byron's famous lines out of "Manfred".' Before he can finish the lines, however, the narrator tells us that:

> he remembered that if nocturnal meditation thereabouts was the fruit of a rich literary culture it was nonetheless deprecated by medical science. The air of other ages surrounded one; but the air of other ages, coldly analysed, was no better than a villainous miasma.[26]

As described already, many of the essays included here challenge the view that 'literary culture' is in some sense 'deprecated by medical science', as James's narrator suggests. Indeed, the final essay in the book's first section celebrates the relationship between science and literature by reading the ways that a number of contemporary British and American poets have engaged positively and effectively with various sciences, technologies, and methodologies. 'Poetry has a responsibility to the *scientia* at its heart,' Peter Middleton argues, 'and if it attends to this it can also find a way to be scientific in ways that do collaborate imaginatively with the sciences of today.'

One of the poets discussed by Middleton in his elaboration of these claims is Allen Fisher, whose poem 'Slow Drag' is included in the second part of this collection. A complex and multi-layered examination of the relationship between science, society, and selfhood, Fisher's poem illustrates what Middleton describes as the value of close observation in this poet's work, but it also engages in a unique way with some of the issues that recur throughout the eight texts included in this part of the book and in the book as a whole – the relationship between self and world, the impact of scientific research on culture and society, the limits of knowledge, and the problem of representing what it is we think we know. 'All engagements of consciousness / involve an aesthetic component,' Fisher writes, reinforcing an idea explored in many of the essays in this collection, but his poem also draws attention to the subjective and social tensions and difficulties that must not be overlooked in detailing the history of science's impact on human affairs. Contributions by poets Iggy McGovern and Randolph Healy explore the various ways that science and literature interact and enrich each other, and both writers include reflections on their own work and the work of others. The second part of the book also includes contributions that explore various aspects of science, from the Aristotelian desire to categorize, describe, and name the things of the world

26 Henry James, 'Daisy Miller' in *Selected tales* (London, 2001), pp 69–70.

in Meredith Quartermain and Maurice Scully's excerpts from long works in progress, to Harry Clifton's description of the different ways that scientists and artists perceive each other and Dylan Harris's replication of the computer programming language C++ in excerpts from a long poem of that title. The collection concludes with Kit Fryatt's poem '*dorade*', a poem that celebrates various forms of discovery – from a child's curiosity about the way things work ('locks // window tax and the plough') to an adult's disbelieving contemplation of 'the Large Magellanic Cloud'.

In his remarkable study of the field known as ''pataphysics' (*sic*) – the pseudoscience invented by the controversial nineteenth-century French writer Alfred Jarry – the poet and scholar Christian Bök has argued that: 'Modern science simply monopolizes the totality of both the subject and the object, leaving no space for poetic wisdom to speak the truth for itself except as an act of deviance within such a norm.'[27] 'Pataphysics is just one area of modern scientific thought left unexplored in this book, but in its presentation of a number of critical and literary voices and viewpoints it might be said to create a space in which 'poetic wisdom' can speak the kind of 'truth' Bök believes is often usurped by scientific discourse in the contemporary world. Equally, it should be said that many writers who have participated in the pursuit of this same 'truth' have not been represented here, from Lucretius to Matthew Arnold, Ralph Waldo Emerson to Marianne Moore, Robert Garioch, Lavinia Greenlaw, and many others. 'Omissions are not accidents', however, as Moore famously said, and the same dictum applies to this book.[28] It is offered to the public in the full knowledge that there is a great deal more to be said about the relationship between literature and science – both of which involve ongoing and open-ended processes of imagination and discovery that keep sending us back to the fundamental questions of why we are here and why the world is as it appears. As the Dutch poet Rutger Kopland has put it: 'in my head someone is looking for words for / something that isn't yet a feeling or a thought'.[29]

27 Christian Bök, *'Pataphysics: the poetics of an imaginary science* (Evanston, IL, 2002), pp 23–4. **28** Marianne Moore, *Complete poems* (London, 1968), p. vii. **29** Rutger Kopland, 'Orchard' ['Boomgard'] in *What Water Left Behind*, trans. Willem Groenewegen (Dublin, 2005), p. 89. Kopland is Professor of Biological Psychiatry at the University of Groningen, the Netherlands.

Part I

Chaucer, technology and the rise of science fiction in English

HELEN CONRAD-O'BRIAIN

Science fiction is, despite serious attempts to legitimatize it, perhaps still the genre most likely mentioned when people complain about the 'expansion of the canon' in literature departments.[1] Readers or viewers who avoid the genre assume that it is outrageously, even irresponsibly, positive or escapist. Extraordinary settings equate with unreality, unreality with irrelevance. Even its defenders and practitioners often effectively deny it that apparently most necessary activity of modern artistic life, the exaltation or at least preoccupation with the individual, an almost automatic forfeit of any claim to literary merit, insisting it is a literature of large concepts as if they are somehow antithetical.[2] While science fiction is idea driven, and characterization is often flattened because the human protagonist stands for a sort of everyman before the problems of a harsh universe and the possibly even harsher difficulties of his or her own nature, this is not

1 Science fiction, like Caesar's Gaul, is divided into three parts, which the reader will need to keep in mind: 1. 'Soft' science fiction based on speculation and extrapolation in the so-called 'soft sciences' (sociology, psychology, and so on); 2. 'hard' science fiction, which is based on speculation or extrapolation from the so-called hard sciences (mathematics, physics, chemistry, astronomy); 3. science fiction of epic and romance (sometimes disparagingly referred to as 'space operas') where the setting is used to update either earlier genre with all its concomitant motifs. It is difficult to keep these three categories separate in practice. 'Soft' and 'hard' in science fiction has nothing to do with sentimentality or its lack. Perhaps one of the most rigorous, unsentimental visions of childhood in English, Henry Kuttner and C.L. Moore's short story, 'Mimsy Were the Borogoves' is essentially 'soft', based on theories of education and child psychology. The classic example of 'hard science' fiction, Tom Godwin's 'The Cold Equations', is an almost unspeakably sentimental work. 2 David G. Hartwell, 'Hard Science Fiction' in David G. Hartwell and Kathryn Cramer (eds), *The ascent of wonder: the evolution of hard science fiction* (London, 1994), p. 35: 'The external universe in all s[cience] f[iction] is distanced from the here and now in part to emphasize that this fiction is not about the specific human condition of any individual today, but about a bigger wider view of all humanity. Robert A. Heinlein plays sophisticated games with ordinary people in "It's Great to be Back", as does Gregory Benford in "Relativistic Effects". But the characters in both stories are relatively flat (versus rounded) and represent or symbolize ordinary people, Everyman. Thus the human condition in hypothetical circumstances is illuminated. It is frequently a surprising literary experience, and one of the ordinary pleasures of all sf.' In the next paragraph Hartwell retreats to a certain extent from this position, but continues to stress the importance of the 'Everyman' in science fiction.

necessary to the genre.[3] The extraordinary and the unreal are not necessarily a retreat from the world around us. Rather they are vehicles to express fears and solutions almost too relevant to a society by transposing them into either an extrapolated future where the problems have come to a head and become clear cut, or by removing them to some other setting which creates a suspension of the reader's immediate prejudices.[4] It is a genre not of observation, in other words, but of postulation. It is the most fictional of all fictions, but it forces maker and reader to face the reality of the human soul. Science fiction does pay more attention to ideas (although not necessarily less to characterization) and, more particularly, to problems modern mainstream fiction does not seem able to accommodate: the implications of ecological change or damage, of technology which seems to increasingly free us from mortality and morality, and the fate of our species within an evolutionary system.[5] Rather than escapist it is above all the genre of the fearful and the problematic.[6]

One method of dignifying a genre is as old as the theory of genres itself – the demonstration of an impeccable (and necessarily ancient) intellectual and artistic lineage. It gives a genre cachet when we can demonstrate the Romantics, Renaissance writers or, best of all, the Greeks indulged in it. This traditional academic sport may seem superfluous, but here such textual archaeology is necessary not merely to understand the development of a genre. The search for antecedents and proto-science fictions should recognize the development of science fiction as a necessary part of tracing how we come to be where we are in our attitudes

3 George Gaylord Simpson, one of the finest palaeontologists of his generation, chose the intersection of three classic science fiction plot lines – time travel, the lost world, and the marooned individual – to write in the novella *The Dechronization of Sam Magruder* a brilliant fiction of the age of the dinosaurs and a character study which his daughter suggested was consciously or unconsciously a transposition of his own interior life, both intellectual and emotional. See Joan Simpson Burns, 'A Memoir' in Joan Simpson Burns (ed.), *The Dechronization of Sam Magruder*, with an introduction by Arthur C. Clarke and an afterword by Stephen Jay Gould (New York, 1996), pp 130–1. 4 Particularly in the most popular of forms, television, science fiction has led public opinion in the US consistently as 'serious' drama could never have and survived. 5 The importance of explaining how things work in any detail has arguably waned in science fiction. The introduction of such explanations suggests the collateral relationship of science fiction to the improving popular scientific writing of the eighteenth and nineteenth century satirized by M.R. James in the opening of 'An Evening's Entertainment'. Jules Verne's work in particular is close to this tradition. See Adam Roberts, *Science fiction* (London, 2000), p. 60. Against my view, see David G. Hartwell, 'Hard Science Fiction' in David G. Hartwell and Kathryn Cramer (eds), *The ascent of wonder: the evolution of hard science fiction* (London, 1994), pp 30–5. 6 This is not to say that science fiction cannot make us laugh – even if the laughter is slightly nervous. See Connie Willis, 'To Say Nothing of the Dog' or Terry Bisson, 'They're made out of Meat' in *Bears Discover Fire and other stories* (New York, 1995), pp 34–7. Both have an almost Wildean quality. Against this reading of science fiction see Hartwell, p. 35: 'This is another of the qualities of hard sf that puts it at the core of the sf genre: It is only truly of interest to people with faith in science, faith that knowledge has meaning. Faith tells them that the universe is ultimately knowable and that human problems (the human condition) are solvable through science and technology; that although science can be misused, if used properly it will lead to the improvement of the human condition, implicitly to heaven on earth. This faith in an improvement in the long run (combining images of evolution with the idea of progress) is a kind of bedrock Darwinism that underlies scientific and engineering culture in the western world in this century. Science fiction is one of the most interesting and eloquent expressions of this faith.'

toward science. And it is a tradition that the genre's own scholars do not always recognize. I.O. Evans included Plato and Lucian in his 1966 collection.[7] The editor of a bibliography on the subject wrote of science fiction's 'very beginning as a literary genre in 1926'[8] and a well-regarded collection of essays with the subtitle *Essays on early science fiction and its precursors* mentions Lucian and a few sixteenth-century texts, but begins its in-depth criticism with the eighteenth century.[9] More recently, the author of the New Critical Idiom's *Science Fiction*, after presenting the ancients versus modern theories of its origins writes: 'I do consider SF to be modern, not ancient; and by modern I mean 'post-Romantic', that is coming after the revaluation of culture and metaphysics associated with the Romantic period, roughly 1780–1830.'[10]

What follows is not, however, the result of looking for science fiction in medieval texts, but of finding it. It was after all the age when the bishop of Paris, faced with the effect of 'curious books' upon the Masters of Arts of the University made belief in the possibility of other worlds very nearly an article of faith.[11] The condemnation of 1277, whatever its intention, encouraged speculation *secundum imaginationem,* which, however alien in some respects to modern approaches to science, still demonstrates a certain kinship with modern science fiction.[12] And when we find Roger Bacon in the later half of the thirteenth century writing like a Latin-language Verne we must look, if only speculatively, at other texts from a slightly different angle:

7 I.O. Evans, *Science fiction through the ages* (London, 1966). **8** Michel Burgess, *Reference guide to science fiction, fantasy and horror* (Englewood, CO, 1992), p. xi. **9** See David Seed (ed.), *Anticipations: essays on early science fiction and its precursors* (Liverpool, 1995). **10** Roberts, *Science fiction*, pp 47–90, p. 54. **11** Edward Grant, *The foundation of modern science in the Middle Ages: their religious, institutional, and intellectual contexts* (Cambridge, 1996), pp 78–9. **12** See Edward Grant, *Physical science in the Middle Ages* (New York, 1971), pp 85–6: 'The condemnation of 1277 and the philosophical and theological consequences that flowed from it in the fourteenth century created an unusual intellectual climate in science and philosophy. No longer was it widely believed that certainty could be acquired about causes and laws of nature. It was now a matter of choosing the most probable of a number of alternatives. Even those who believed inwardly that acquisition of scientific truth was possible – usually masters of arts – were forced by the change of attitude to couch their conclusions in hypothetical language. A sophisticated positivistic attitude developed in the fourteenth century. Mertonians and Parisians, who contributed most to scientific thought, abandoned hope of acquiring true knowledge of the physical world. Instead they channelled their energies into hypothetical discussions such as intension and remission of forms; or they conjured up fascinating hypothetical problems, as, for example, the way bodies would behave if placed in a vacuum that God had created between the moon and the earth. […] They were clearly content to exercise their scholastic ingenuity on hypothetical problems "according to the imagination" (*secundum imaginationem*). Solutions to such problems were not intended for application to nature. Logical consistency, not a quest for physical reality, was the major objective. Certain of the topics discussed in this context, and the conclusions and proposed solutions that emerged, were destined to play a vital role as the scientific revolution unfolded in the sixteenth and seventeenth centuries. Examples are the discussions on the diurnal rotation of the earth, the kinematic theorems embedded in treatises on intension and remission of forms, and the investigations concerning hypothetical vacua. But these ingenious conclusions and solutions could play no role in the development of a new science until they were divorced from an approach to nature characterized by *secundum imaginationem* and "saving the phenomena" and associated instead with a quest for physical reality.'

> Therefore first I will tell of works of art and ingenuity and marvels of nature that afterwards I may speak in detail of their advantages and construction in which there is nothing magical, so that it may be seen that all magical powers be inferior to these works and worthless. And first I will speak of their shaping and planning by art alone. For sailing machines are possible without human governance like the biggest ships for the river and the sea, run by a single man, with such speed as if they were full of men. Again chariots are possible that can move at an incalculable speed without animals as we suspect the scythed chariots were which were used in ancient time for battle. Again it is possible to construct a flying machine so that a man seated in the middle of the machine revolving some sort of machinery by means of artificial wings beats the air as birds do. Again it is possible to build a machine, small in size to raise and lower weights of almost infinite size [...] There are able to be machines for moving in the sea or in rivers right at the bottom without any physical danger. For Alexander used such a machine and saw the secrets of the sea according to what Aethicus Ister wrote. These things have been made in ancient times and in our own times as is manifest, except the flying machine which I have not seen nor any man I know has seen, but I know a wise man who knew how to build such a machine.[13]

However marvelous these machines are, they are not magic. Compared to them magic is *inferior* and *indignus*, 'inferior and unworthy'. They are machines built either in his own day or in antiquity, except for the *instrumenta volandi* which draws forth a touching half disclaimer.

Perhaps we should at least begin to suspect that sometimes medieval and even earlier literature is like science fiction rather than that science fiction is like it. Perhaps what looks like magic is not necessarily magic, but in fact 'hard' science fiction. Perhaps science fiction may even have as reasonable a generic claim on that masterpiece of characterization, Chaucer's 'Franklin's Tale' as any other form.[14] Let us begin with a few basic assertions:

1 Science fiction has been around for a very long time;
2 Science fiction has been around for a long time because our species is tool-using and has an insatiable urge for knowledge, power and outward expansion – not necessarily in that order;

13 My translation. Roger Bacon, '*De instrumentis artificiosis mirabilibis*', *Epistola Fratrus Rogeris Baconis de secretis operibus artis et naturae, et nullitate magicae*, Cap. iv, ed. J.S. Brewer, Appendix I, Rolls Series 15 (London, 1859), pp 532–3. **14** Against the argument that science fiction is not conducive to character development, we find in 'The Franklin's Tale' all the latest fourteenth-century methodologies and literary tactics of characterization.

3 Science fiction is question and fear driven;
4 The questions and fears we find in science fiction derive from our nature and circumstances as a species.

Science fiction then is not essentially about any particular technology or natural law. It is about extraordinary power and the unknown entering human society through knowledge. At an earlier point in the English language 'can' meant both 'to be able' and 'to know'. It is a pity we have lost the double meaning: it expresses perfectly what this power and knowing involves. Sometimes science fiction expresses this in terms of humans who know and can do things other humans do not know and cannot do. Sometimes it expresses both by beings who are like, but unlike us, who 'can' what we cannot. Indeed, science fiction and tales of magic and faerie share two characteristic 'plot' problems. In the first a human gains or seeks power through knowledge and/or technology and struggles or succumbs to the concomitant temptations and moral dilemmas for flawed humanity. In the second humans individually or collectively face beings with powers, values and conduct which threaten or may threaten the individual and society. These problems generate narrative motifs and fuel the modern science fiction narrative engine as they once did that of the Breton lai. They feed on the same human desire for, and dangerous fascination with, the power to override the normal processes and limitations of nature. Faust damns himself for power and Prospero must break his staff and renounce his books to take his place again in human society.

These stories, however we define their genres, resort, again and again, to the equally human fear of the Other. This Other is more likely to be represented as superior or inferior than as equal,[15] to represent, in any given text, the latent or possible inhumanity within ourselves, triggered by an access of power (and knowledge is power in science fiction as much as it is in fairy tales) that removes us from the normal restraints of our humanity and by extension, of our morality. Possession of superhuman power, whether innate or technological, whether by the extra-terrestrial or the more traditional inhabitants of Faerie (who have contributed so much to the popular paradigm of the extraterrestrial),[16] generates

15 It is possible to reverse these situations and explore the Other as less powerful than ourselves, a plot device which the writers of the four *Star Trek* series and movies largely denied themselves by the invention of the 'Prime Directive', with some notable derogations: 'A Piece of the Action', 'Homecoming', 'Who Watches the Watchers', and 'A Private Little War'. **16** On this point see Diane Purkiss, *Troublesome things: a history of fairies and fairy stories* (London, 2000), p. 320. Katherine Briggs has drawn attention to an apposite development in at least some later medieval romances: the 'fairy lady' is not another being, but a learned woman, giving an example from the French prose *Lancelot*: 'In those days all maidens that knew enchantments or charms were called fays, and there were many of them at this time, and more in Great Britain than in other lands. They knew, as the story saith, the virtue of herbs and of stones and of charms, whereby they might be kept in plenty and in the great wealth that they possessed.' See *Sir Lancelot of the Lake: a French prose romance of the thirteenth century.* Trans. Lucy Allen Paton (London, 1929), p. 6, quoted in Briggs, *The fairies in tradition and literature* (London, 1967), p. 5.

much of this otherness and objectifies and excuses the fear which derives from the perceived or imagined differences. The faerie mistress, to take a common example, is clearly a female, usually an unusually desirable female from the standpoint of the human male, but one whose aberrant behaviour (in a human context) springs, at least in part, from her particular superhuman powers which are an integral and defining part of her non-human nature. The prohibition she places on her human consort, symbolic of her place to one side of the human norm, is often one which is chosen to cause little embarrassment to her spouse or lover, but invariably symbolizes her position as Other than human and possessing more than human powers. No daughter of Eve, she has no reason, biblical or otherwise, to be subservient to her human husband, except out of courteous affection. But in *Sir Orfeo*, for example, the often relatively benign intrusion of one faerie woman into human life is replaced by the intrusion of faerie society as a whole in the person of its king. Entering almost literally at the bottom of the garden, he disrupts human society from the basic unit of marriage through the association of human beings in nations/kingdoms to the normal relationships between sovereign states.[17]

Many anthologies of science fiction begin with Edgar Allan Poe, or with Mary Shelley's *Frankenstein*, or with *The Life and Adventures of Peter Wilkin,* but science fiction really begins where myth ends. At certain moments, they almost seem to co-exist. That Apollo is the 'far shooter' and the Anglo-Saxons spoke of 'elf-shot' suggest a time when the mechanics of the bow were still, to adapt Clarke, sufficiently advanced to appear not merely magical but divine. Technology has always had a whiff of the superhuman/supernatural about it. Chaucer, himself, in 'The Canon's Yeoman's Tale' makes his narrator speak of 'oure elvysshe craft' (VII [G] 751) and 'this eluyssche nyce loore' (VII [G] 842).[18] But even before Lucian imagined the heavens as mundane in his ethnographical burlesques, Daedalus had flown out of peril into grief and Homer had introduced automatons in lines 417–20 of Book 18 of the *Iliad*:

> And in support of their master moved his attendants.
> These are golden, and in appearance like living young women.
> There is intelligence in their hearts, and there is speech in them
> And strength, and from the immortal gods they have learned how to do things.[19]

17 See Helen Conrad-O'Briain, 'Some reflections on *Sir Orfeo*'s Mirror of Polity', in Anne-Marie D'Arcy and Alan J. Fletcher (eds), *Studies in late medieval and early Renaissance texts in honour of John Scattergood* (Dublin, 2005), pp 76–90. **18** Unless otherwise indicated, all references to the text of *The Canterbury Tales* are to Larry D. Benson (ed.), *The Riverside Chaucer* (3rd ed. New York, 1987). **19** *The Iliad of Homer*, trans. Richmond Latimore (Chicago, 1961) p. 386.

Evans wrote in the introduction to his collection of the earliest science fiction that:

> Science fiction is not, as is sometimes thought, a recent innovation: it is the modern development of that age-old art-form the 'wonder story'. [...] Such narratives, however fantastic they may be, base themselves on contemporary culture, so it is not surprising that the wonder story of the Technical Age should base itself on technics. [...] The essential difference between the old-time wonder story and science fiction is that the former is based on magic and the latter on what at any rate purports to be science.[20]

Evans apparently overlooked how early technology does enter these stories – how soon magic begins to be rationalized. Homer has already brought the wonderful out of the necessary bondage of magic. The mind has traveled a long way in its acceptance of the power of the made thing when a god manifests his power in his control of technology. The limping god Hephaestus builds automatons to help him at his forge. They are clearly *ingenium,* machines built to help him overcome the limits of his own 'body'. It is a landmark in Western engagement with and understanding of technology. Once magic or divine power begins to be explained in terms of the manipulation or understanding of natural law and technology, science fiction is at the door, if not in the window. The wizard or supernatural being becomes the scientist, mad, benign, or otherwise – Shakespeare's *The Tempest* becomes Irving Block and Allen Adler's *Forbidden Planet.*

The appearance and reappearance of science fiction is inextricably linked to technological and scientific change. Hephaestus' automatons and Alexander's submarine seem connected with the growth of Greek mechanics. Medieval examples may be connected to the technology and science introduced to the West by the translations of the twelfth and thirteenth centuries which increasingly moved out of the universities into the horizon of the upper and scribbling classes. The passage from Roger Bacon quoted above reminds us that Alexandrian engineering had not been lost or even mislaid – it had simply lost its audience (it never had – except perhaps in the besieged Syracuse – a real market). However marvelous these things might seem to the medieval audience, Bacon treated them as constructions, possible and practical.[21] It could even be argued the passage reflects a strong and optimistic sense of the progress of engineering on the author and his readers, whose effects they could see around them, particularly in architecture and textiles.[22]

20 Evans, pp 9–10. **21** The word 'marvellous' must be stressed here, for it is exactly this quality rather than practical application that appealed to the engineers of the Caliphate. See, for example, Donald R. Hill, 'The Banù Mùsà and their *Book of ingenious devices*', *History of Technology*, 2 (1977), 39–76, reprinted in *Medieval Islamic technology: from Philo to al-Jazarì – from Alexandria to Diyàr Bakr*, ed. David A. King (Aldershot, 1998). **22** The origins, spread and importance of the horizontal treadle loom and the spinning wheel have

Modern readers may boggle at the science from which this medieval science fiction springs, particularly if their definition of science fiction is essentially that of 'hard' science fiction. Any narrative relying on 'hard science' written prior to the scientific revolution may immediately be relegated to fantasy. Magic, however, may be merely science we do not accept. Much or most of Verne, even when he is most careful of his technology, is patently impossible, but no one suggests that it is not science fiction for that reason.[23] It would be perfectly valid in the fourteenth century to write science fiction based on the Ptolemaic system or using the theories of optics then available. Science fiction would not be any less science fiction if used theories of time expressed in the writings of a scholastic philosopher like Robert Kilwardby[24] to create and maintain the plot. To be science fiction, I would suggest, the narrative must be congruent to what the writer and the original audience believed to be possible given possible technologies and/or the reality of nature, what medieval scholars would have called 'saving the appearances' or 'saving the phenomena.' The real problem the modern reader must face in assigning a work like Chaucer's 'Franklin's Tale' to the genre is one of definition, and not of the genre, but of the science, and in particular 'magyk natureel', 'sciences by whiche men make diverse apparences' (*Canterbury Tales*, Group F: ll.1139–40). The distinction, as Spearing wrote, is in the treatment.[25]

A modern reader looking for science fiction in Chaucer would be forgiven for assuming the place to begin would be the 'Canon's Yeoman's Tale'. If the tale were actually about alchemy, that picturesque parent of chemistry, we might have some reason for looking at it in detail, but alchemy as alchemy plays very little part in the narrative. The 'Canon's Yeoman's Tale' lies within the picaresque tradition with practically no overlap with science fiction. It may be the ancestor of *The Sting*, but it would be difficult to see it in a direct line of descent to *2001*. Something quite different emerges, however, when we turn to 'The Franklin's Tale'. I would suggest that the first hard science fiction in English – if we take hard science fiction to be a narrative in which the physical laws and the application of those laws, with or without a concomitant technological application,

yet to be properly understood and appreciated. See the brief, but trenchant, remarks in Richard Holt, 'Mechanization and the medieval economy' in *Technology and resource use in medieval Europe: cathedrals, mills and mines*, ed. Elizabeth Bradford Smith and Michael Wolfe (Aldershot, 1997), pp 139–57. Lynn White's optimistic and influential view of powered machinery in the early Middle Ages has been effectively challenged, but the development of technology, not necessarily its widespread use is at issue here. (See L. White, Jr, *Medieval technology and social change*, [Oxford, 1962].) For reevaluations of White, see Holt, p. 141, and Michel Toch, 'Agricultural progress and agricultural technology in Medieval Germany: an alternative model' in Smith and Wolfe, pp 158–69. **23** See M. Hammerton, 'Verne's amazing journeys' in Seed, pp 98–110. **24** See Robert Kilwardby, *On time and the imagination: de tempore de spiritu fantastico*, ed. P. Osmund Lewry, Auctores Britannici Medii Aevi IX (Oxford, 1987), pp 7–44. An English translation is available in Robert Kilwardby, *On time and the imagination, part 2: introduction and translation*, ed. Alexander Broadie, Auctores Britannici Medii Aevi IX (2), (Oxford, 1993), pp 25–63. **25** Geoffrey Chaucer, *The Franklin's Prologue and Tale*, ed. A.C. Spearing (Cambridge, 1994), pp 61–5. Note in particular Chaucer's comments on illusion.

directs the narrative to a significant effect – is Chaucer's 'Franklin's Tale'. In it the ability of the Clerk of Orleans to predict or bring about a natural phenomenon, based on a knowledge of natural laws, is as important to the narrative as the laws of physics are to Tom Godwin's science fiction story 'The Cold Equations' (1954). Nevertheless, it would be wrong to dismiss 'The Canon's Yeoman's Tale' out of hand in an attempt to demonstrate the existence of science fiction in *The Canterbury Tales*. Among the rich diversity of Chaucer's interests, welling up like a turbulent spring in his writing, we cannot avoid the scientific, and it is well to remind ourselves of it, from *A Treatise on the Astrolabe* to 'The Canon's Yeoman's Tale' itself. Consider the Canon's come-on to his 'mark'. Chaucer does not have him say: 'Taak in thyn hand, and put thyself therinne / Of this quyksilver an ounce, and heer bigynne, / In name of Crist, to wexe a *magiciene*.' Rather, he puts these words in his mouth:

> Taak in thyn hand, and put thyself therinne
> Of this quyksilver an ounce, and heer bigynne,
> In name of Crist, to wexe a *philosofre*.
> [..............................]
> For ye shul seen heer, by experience
> That this quyksilver I wol mortifye
> Right in youe sighte anon, withouten lye,
> and make it good silver and as fyn
> As ther is any in youre purs or myn
>
> (VIII [G] 1120–2, 1125–9; emphasis added)

The Canon for all his rogue's patter uses no hocus pocus, no invocation of spirits here; quite the opposite – 'in name of Crist'! His method, however bogus, is bogus science not bogus magic: the addition of heat and 'a poudre'. We know that 'poudre' is worthless, 'Ymad, outher of chalk, outher of glas' (l. 1149), but the method is neither magical nor dressed up as magic. It is merely the reaction of material and heat upon other materials to create the circumstances to transform quick into true silver. In 'The Squire's Tale', immediately preceding 'The Franklin's Tale', the story hinges on marvellous objects, and magic hangs heavily over them, but not so as to completely obscure the possibility of technology. The description of one of them, a mechanical flying horse controlled 'with writhing of a pyn' (V [F] 127), includes insinuations of natural magic in the mention of constellations and seals:

> He that it wroghte koude many a gyn.
> He wayted many a constellacioun
> Er he had doon this operacion,

And knew ful many a seel and many a bond. (V [F] 127–30)[26]

While the foundation text on the technology of automata did not reach Western Europe until the sixteenth century, the existence of such 'gyns' was known from antiquity.

'The Franklin's Tale', described by its narrator as a 'Breton lai', is a spider's web of themes. Two, however, do stand out: self-respect based on adherence to an ethical ideal, and the corruptive nature of power. No right-minded fourteenth-century audience would recognize what Aurelius experiences as anything but a disordered passion. However much he claims to love her he desires not the good of the beloved, but rather he desires possession of Dorigen, power over her. To have this power he is willing (at least in the abstract) to do whatever is necessary: to lie, to cheat, to squander his patrimony. He will employ whatever means are available; it is here that we and Chaucer arguably enter the purview of science fiction, for when the traditional arts of persuasion and the pagan gods fail him,[27] he resorts after giving himself up to the expected lover's complaints to science in the person of the Clerk of Orleans. I think here that we should suspend the identification of Orleans and its 'yonge clerkes, that been lykerous / To redden artes that been curious' (V [F] 1119–20) with Merlin and Morgan le Fay.[28] The knowledge and skills that the young clerk of Orleans brings to bear on the problem of removing the rocks from sight, and let us remember that is all he is able or has been asked to do, are described in a way that does not recall magic so much as science. The Clerk of Orleans with his power at the command of another is not a malevolent sorcerer or a sorcerer of any kind, rather he is an ancestor of the scientist as innocent in an evil world, or morally neutral in an ironically thoughtless pursuit of knowledge that can be used for good or evil.

On a deep level 'The Franklin's Tale' is about natural law as the Middle Ages understood it – the unchanging realities at the foundation of creation, a law which embraced both the physical order of the universe and the moral order which according to theologians and jurists were shared by all peoples regardless of religion. Setting the stage for the action which follows, Dorigen questions the purpose of the rocks which have come to embody all her fears for her husband:

26 In a note on lines 115–342 in *The Riverside Chaucer*, it is suggested that '[t]he magic stede of bras is described primarily as a mechanical aeronautical contrivance controlled by the manipulation of various pegs.' (p. 892) In other words, it is no mere delusion. **27** Although many medieval men and women would assume that anything to do with natural magic had by force to include the implicit intervention of demons who were in effect the 'gods of the pagans'. See, for example, the passage from the *Summa Theologiae* quoted by Gerald Morgan in the notes to his edition of 'The Franklin's Tale' (Dublin, 1988), p. 101. **28** Although Thomas Malory writes of that lady 'And the third syster, Morgan le Fey, was put to scole ina nunnery, and ther she lerned so moche that she was a grete clerke of nygromancye'. Eugène Vinaver (ed.), *Works* (2nd ed. Oxford, 1971), p. 5. See also note 16 above.

Eterne God, that thurgh thy purveiaunce
Ledest the world by certein governaunce,
In ydel, as men seyn, ye no thing make.
[. . . . .]
Why han ye wroght this werk unresonable? (V [F] 865–7, 872)

Later she will rely on the natural and unchanging reality of the physical world to rid her of Aurelius's unwelcome advances, 'For wel I woot that it shal never bityde.' (V [F] 1001).[29] For his part, even in his prayer to Apollo, Aurelius, although he twice uses the word 'miracle' (V [F] 1056, 1065), and suggests as an alternative that Diana 'synken every rok adoun / into hire owene dirke regioun' (V [F] b 1073–4) thinks and speaks in scientific terms: the passage is like a layman's echo of the hypothesis *secundum imaginationem* which busied more than simple Masters of Arts in Chaucer's century. He is represented as sensitized to the possibility of help, which is not perhaps as supernatural as it at first appears:

Ye know en wel, lord, that right as hir desir
Is to be quyked and lighted of youre fir,
For which she folweth yow ful bisily,
Right so the see desireth naturelly
To folwen hire as she that is goddesse
Bothe in the see and ryveres moore and lesse.
Wherefore, lord Phebus, this is my requeste –
Do this miracle, or do myn herte breste –
That now next at this opposicion
Which in the signe shal be of the Leon,
As preieth hire so greet a flood to brynge
That fyve fadme at leeste it oversprynge
The hyeste rokke in Armorik Briteyne;
[. . . . .]
Preye hire she go no faster cours than ye;
Thanne shal she been evene ate fulle always,
And spryng flood laste bothe nyght and day
(V [F] 1049–61, 1068–1070)[30]

We are used to science fiction being set in the future – unless it is in an Atlantean sub-genre. Here we see it thrown into the past, but not into a lost civilization, although an important component of fourteenth-century science and technology

29 While there is a tradition of sunken lands around the coasts of Britain and France, this was not allowed to impinge on Dorigen's consciousness. **30** '[T]he mean range of spring tides locally is such that if the flood tide were this high, the ebb tide would just cover the rocks.' Spearing, p. 14.

was the western recovery of Alexandrian science then over a thousand years old. There is nothing to suggest the Franklin tells a story in which the frisson of 'lost knowledge' plays a part. They are 'curious arts' but the Franklin does not actually suggest that there is anything diabolical about them.[31] He assures his audience that only in a pre-Christian past could such things occur:

> Which book spak muchel of the operaciouns
> Touchynge the eighte and twenty mansiouns
> That longen to the moone, and swich folye
> As in oure dayes is nat worth a flye, –
> For hooly chirches feith in oure bileve
> Ne suffreth noon illusioun us to greve. (V [F] 1128–1134)

The Church now stands between its children and the deceits of such illusions. But how and why? The Church would not and could not ignore the 'eighte and twenty mansiouns / That longen to the moone' in the search for an improved calendar and computus.[32] It employed and supported astronomy. It has not denied the faithful the study of the natural operations of nature, but has freed the 'illusions' whether practiced by demons or by clerks, by diabolic agency or natural magic, by unmasking their source and nature.[33] The original audience may have been aware of a more technical reason for the setting. Oresme, an older contemporary of Chaucer,[34] questioned the usefulness of astrology at a number of points, one of which may have a bearing here:

> The second part is a part of natural science and is a great science and it too can be known insofar as its nature is concerned but we know too little about it and in particular the rules of the books are false, as Averroes says, and have slight proof or none. And some of them were fulfilled in the place or at the time when they were laid down, are false in other places or at the present time: for the fixed stars which according to the ancients have great influence are not now in the position they were in then and these same positions are used in making predictions.[35]

31 Compare the methods used by Tebano in Boccaccio's *Il filocolo,* a source for Chaucer's tale: *Chaucer's Boccaccio: sources of Troilus, and the Knight's and Franklin's tale*, ed. and trans. N.R. Havely (Woodbridge, 1980), p. 157. **32** Wesley M. Stevens, 'Cycles of time: calendrical and astronomical reckonings in early science' in J.T. Fraser and L. Rowell (eds), *Time and process: the study of time* VII (Madison, 1993), pp 27–51. **33** It would be interesting to know when the concept of the fairy illusion enters the tradition. There is no sign of it, for example in *Sir Orfeo.* Is it a 'learned' theological concept which has infiltrated 'fairy lore'? **34** He was born *c.*1320 near Caen and died in 1382 as bishop of Liseux. See Edward Grant, 'Scientific thought in fourteenth century Paris: Jean Buridan and Nicholas Oresme' in *Machaut's world: science and art in the fourteenth century*, eds. Madeleine Pelner and Bruce Chandler (New York, 1978), pp 105–24. **35** Edward Grant (ed.), *A source book in medieval science* (Cambridge, MA, 1974), p. 489

Ironically, Oresme's influence was limited not only because of the technical difficulties posed by much of his work, but by his attacks on natural magic and astrology.[36]

The introduction of power through a technology or knowledge of natural laws, here astronomy, refreshes the old narrative categories. A new genre has appropriated the maiden's rash promise. What might have been a Faustian narrative has turned into something quite different. This is a point of particular interest. Despite the essential objections to natural magic, that it blasphemes in effect against God's providence and calls upon demons to create effects, there is no allusion to the diabolic. The reference to God's providence is by extension and implication possibly because of its pre-Christian setting. The character, the Clerk, supplying the power is no longer a morally suspect Other, but a human being with free will, compassion and, more importantly, the desire to be well-considered. The person who employs him does not pledge his soul, but rather his financial well-being and social position – the circumstances which would not be out of place in a nineteenth-century novel. The basic narrative could have been easily driven by pure magic rather than by the misuse of a science, astronomy. Aurelius could have found a wizard – or met the devil roaming about looking for his prey, either of whom could have removed the rocks (whether actually or not). There is something suspiciously modern about this tale. Just as Wilkie Collins' *The Moonstone* bathes in the atmosphere of the mysterious East, while the true mainspring of the action is the nineteenth-century recognition of the importance of the inner workings of the subconscious mind.

It is true, that Aurelius and his brother come upon a clerk, apparently loitering with intent to commit magic, who after greeting them 'seyde a wonder thing: / 'I knowe […] the cause of youre coming.' (1175–6) Possibly some of Chaucer's audience may have identified the clerk's pre-dinner entertainments, undoubtedly to establish his powers with the brothers, illusion, as the product of some sort of advanced camera obscura. Some of them, attempting to work the necessary arrangements, may have decided the poet had seriously outrun the possibilities of technology, but so too, we are told, do writers who require faster than light or time travel, but no one suggests that even the *Enterprise* or the Tardis work by magic. Modern readers may immediately dismiss any suggestion that in 'The Franklin's Tale' we are reading science fiction because of the Franklin's use of the term 'magyk natureel', but this magic 'science by whiche men make diverse apparences' (V [F]1139–40) is not, in more ways than one,

36 Grant, 'Scientific Thought', p. 107. His remarks on astrology are not, however, inappropriate in connection with Chaucer. His close connections with the courts of Jean II and Charles V may have brought him 'into personal contact with Machaut, a suggestion that gains in plausibility from Oresme's evident interest in music, and his apparent acquaintance with Philippe de Vitry, the great champion of the *ars nova* in the first half of the fourteenth century.' Grant, p. 106 and n. 7.

what it seems. In the notes to his edition, Gerald Morgan has rightly marshalled texts from Aquinas opposing 'magyk natureel' on theological and philosophical grounds,[37] but while it can be called 'bad science' on more than one level, it cannot be denied it was approached as science, and continued to be approached as science into the century of the scientific revolution, even gaining momentum when it was taken up by Ficino. As much as he disparaged it along with astrology and alchemy 'which have had better intelligence and confederacy with the imagination of man than with his reason', Bacon wanted at least to retain the name for experimental science since 'natural magic pretendeth to call and reduce natural philosophy from variety of speculations to the magnitude of works.'[38] Magic in the treatises translated from Arabic, or deriving from those treatises, during the twelfth and thirteenth centuries, included a branch which used 'talismans' made without the invocation of or adjuration of spirits, but only by inspecting the state of higher bodies. This, according to Adelard of Bath, was the culmination of the liberal arts.[39] It is also arguably the curious art which Aurelius's clerk uses on his behalf. Aquinas's own master, Albertus Magnus, attempted in his *Speculum astronimiae*, 'to sort out which books in this area were licit and illicit' and could not bring himself to destroy even the most suspect works outright for the sake of utility.[40]

As the Franklin describes it, however, it is as confused as it is bad and trivial besides. Terms like 'illusioun' and 'japes' recall Oresme's own attack on astrology in the *Livre de Divinacions*. Oresme, as we have seen above, cast serious doubt on the basis of actual knowledge from which the rules of astronomy were derived.

> The first part of astrology is speculative and mathematical, a very noble and excellent science [...] and this part can be adequately known but it cannot be known precisely and with punctual exactness, as I have shown in my treatise on the Measurement of the Movement of the Heavens and have proved by reason founded on mathematical demonstration. [...] [A]s regards change in the weather, this part by its nature permits of knowledge being acquired therein but it is very difficult and is not now, nor has it ever been to anyone who studied it, more than worthless[41]

37 See Morgan (1988), pp 101–3, and especially p. 103: *Summa Theologica* 2a2ae96.1corp: *Est etiam hujusmodi ars inefficax ad scientiam acquirendam. Cum enim per hujus modi artem non intendaturacquisitio scientiae per modum homini connaturalem, scilicet adinveniendo, vel addiscendo, consequens est quod iste effectus vel expectetur a Deo, vel a daemonibus.* ('It is also useless for the advance of science. It does not follow the method natural to man, that is by research and instruction, but expects information to come either from God or from the demonic.' My translation.) **38** Francis Bacon, *The advancement of learning and the new Atlantis*, ed. Arthur Johnston (Oxford, 1974) p. 31. See also pp 88–97. **39** See Charles Burnett, 'Talismans: magic as science? necromancy among the seven liberal arts' in *Magic and divination in the Middle Ages: texts and techniques in the Islamic and Christian worlds* (Aldershot, 1996), pp 1–15. **40** Morgan (1988), p. 3, pp 13–15. **41** Grant (ed.), *A source book*, p. 489.

There is no question here of mighty works which would draw down the power of celestial power for good or ill by the manipulation of the lower. It is bad science because it is of necessity inexact. That it is 'bad' science both ethically and practically makes it the perfect engine for creating the circumstance in which Dorigen finds herself. Corrupt, it is an agent of corruption, working by means of illusion. No one with good intentions would want to use such trickery. There is, however, something particularly curious about these arts as the Franklin uses them in his story. He takes much from astronomy, something by implication from medical astrology, and calls the lot 'magic natureel' which might, presented only slightly differently, be better called astrological meteorology. The treatment of the clerk's methods seems to stress the idea that science hopelessly embrangles, morally as well as technically, in an elaborate astronomical fraud – a 'sting' on one level not so different as that perpetrated on the priest in 'The Canon's Yeoman's Tale':

> To wayten a tyme of his conclusioun;
> That is to say to maken illusioun,
> By swich an apparence of jogelrye –
> I ne kan no termes of astrologye –
> That she and every wight sholde wene and seye
> That of Britaigne the rokkes were aweye,
> Or ellis they were sonken under grounde.
> So atte laste he hathn his tyme yfounde
> To maken his japes and his wreccednesse.
> Of swiche a supersticious cursednesse
> His table Tolltanes forth he brought
> Ful wel corrected, ne ther lakked nought
> Neither his collect ne his expans yeeris,
> Ne his rootes, ne his othere geeris,
> As been his centris and his argumentz ... (V [F] 1263–77)

I cannot find any suggestion that such astrological talismans could be used to create illusions. They could be used medically, or to gain the affection of a ruler, or to regain that of a wife or husband.[42] Of the actual method used to create the illusion, however, the audience is given not a clue beyond the mention of 'geeris', equipment, which might be the suspect talismans otherwise in Chaucer connected with medical astrology, or simply the equipment of astronomy, and what he calls 'observaunces / For swiche illusiouns and swiche meschaunces / As hethen folk useden in thilke dayes.' (V [F] 1291–3) As long as we are in the realm of real science, discovering when certain celestial conjunctions will occur, the description

42 Burnett, p. 11.

runs on and on preparing us, blinded by science, to accept, of all things, the reality of the clerk's illusion. The point where the hard science ends and the theories begin, the Franklin and Chaucer fall silent. Why? The Franklin 'kan' indeed 'termes of astrologye.' Does the clerk use 'ymages' as in 'The House of Fame' to focus the influence of the planets? Chaucer writes:

> And clerkes eke, which konne well
> Al this magik naturel,
> That craftily doon her ententes
> To make, in certeyn ascendentes,
> Ymages, lo thrugh which magik
> To make a man ben hool or syk. (ll.1265–70)

Although there, as in the description of the Doctor of Phisik in the 'General Prologue' (ll. 414–18), the use is medicinal rather than meteorological or oceanographic? Why is all else silence? Does the Franklin find it distasteful to speak of such things? Why then should he tell such a tale? Has Chaucer done what writers of science fiction have always done, to extrapolate from known science and technologies? He has extended the use of talismans in a questionable science to create what is nothing more than an illusion.

Whatever the clerk does, once he has his dates right, he does not really change anything. His science in reality is as helpless before the rocks as ours still is before nature, creating, like the levies of Lake Pontchartrain, only the illusion of safety, an illusion more dangerous than reality. There is something in the way Chaucer shapes the discourse around this attempt at subverting the moral order by an illusionary subverting of the natural order which draws the audience's attention by obfuscation. The tension between the undoubted pull of the moon on the sea and the possible power it might provide, between the technical language, even jargon, of science and the never elaborated 'observances' as the clerk 'pulls the thing off' suggests that Chaucer is using the tension at the heart of science fiction – between the possible and the not necessarily impossible – to intensify the misdirection of human desires and strivings which a moment of human compassion will return to their proper course:

> And in his herte he caughte of this greet routhe,
> Consyderynge the beste on every side,
> That fro his lust yet were hym levere abyde
> Than doon so heigh a cherlyssh wrecchednesse (V [F] 1520–3).

A pocketful of death: horology and literature in Renaissance England

JOHN SCATTERGOOD

My uncle William (now deceased, alas!) used to say that a good horse was a good horse until it had run away once, and that a good watch was a good watch until the repairers got a chance at it.

Mark Twain, 'My Watch – An Instructive Little Tale'[1]

On 27 February 1614 John Harrington, 2nd Baron of Exton, died of smallpox at the age of twenty-two. John Donne, who was a friend of his sister Lucy, countess of Bedford, wrote an elegy for him.[2] The dead young man was said to have been 'the most complete young gentleman of his age this kingdom could afford for religion, learning and courteous behaviour',[3] and in one place Donne compares him to a pocket watch and regrets that the small compass of his life did not allow him to become like a public clock:

> Though as small pocket-clocks, whose every wheele
> Doth each mismotion and distemper feele,
> Whose *hand* gets shaking palsies, and whose *string*
> (His sinews) slackens, and whose *Soule*, the spring,
> Expires, or languishes, whose pulse, the *flye*,
> Either beates not, or beates unevenly,
> Whose voice, the *Bell*, doth rattle, or grow dumbe,
> Or idle,'as men, which to their last houres come,
> If these clockes be not wound, or be wound still,
> Or be not set, or set at every will;
> So youth is easiest to destruction,

1 Mark Twain, *Collected tales, sketches, speeches, & essays, 1852–1890* (New York, 1992), pp 497–9. 2 See Herbert J.C. Grierson (ed.), *Donne: poetical works* (1929; rpt. Oxford, 1979), pp 247–54; p. 251. 3 The words are those of Sir James Whitelocke in *Liber famelicus*, p. 39; quoted in W. Milgate (ed.), *John Donne: the epithalamions, anniversaries and epicedes* (Oxford, 1978), p. 197.

If then wee follow all, we follow none.
Yet, as in great clocks, which in steeples chime,
Plac'd to informe whole towns, to'employ their time,
An error doth more harme, being generall,
When, small clocks faults, only'on the wearer fall;
So worke the faults of age, on which the eye
Of children, servants, or the State relie.
Why wouldst not thou then, which hadst such a soule,
A clock so true, as might the Sunne controule,
And daily hadst from him, who gave it thee,
Instructions, such as it could never be
Disordered, stay here, as a generall
And great Sun-dyall, to have set us All? (ll. 131–54)

The comparison is merely the vehicle for what Donne is trying to say here – which is that Harrington's life was so exemplary that it was a pity that it should have been so short, for, had he lived longer, he would have provided a model of behaviour for others, just as an accurate public clock regulates the behaviour of whole towns, whereas a good pocket watch is a personalized timekeeper and regulates the behaviour of its owner only.

Nevertheless, what the passage says or suggests about clocks, and particularly about pocket watches, is interesting. It is clear that Donne knows about the mechanics of pocket watches: we know that he possessed one, for he bequeathed to his brother-in-law Sir Thomas Grymes 'that striking clock which I ordinarily wear'.[4] All early clocks and the large public clocks of Donne's day were powered by falling weights, but in watches the 'spring' provided the motive force and it is a spring-driven watch that Donne has in mind here. One of the problems with early spring-driven watches was the equalization of motive force as the mainspring wound down. This was addressed first by a stackfeed mechanism, and later by the device of the fusee wheel, conical in shape, which acted as an intermediary wheel between the mainspring and the wheeltrain of the watch, and around which a cord (later a chain) passed.[5] This is the 'string' Donne mentions here and he explains its function in one of his sermons: 'In a watch, the string moves nothing, but yet, it conserves the regularity of the motion of all'.[6] Since this watch has a 'bell' it must have had a striking train, and the 'fly' is the device which regulates the speed of the stroke. The singular 'hand' is interesting, because it suggests that Donne had in mind a watch with only an hour hand –

4 See R.C. Bald, *John Donne: a life* (Oxford, 1970), p. 563. **5** For an elegant and brief explanation of the stackfeed and the fusee wheel in relation to early watches see R.W. Symonds, *A book of English clocks* (London, 1950), pp 20–6. **6** See G.R. Potter and Evelyn M. Simpson (eds), *The sermons of John Donne*, 10 vols (Berkeley, 1953–62), VII, p. 430.

which was the case with most early watches.[7] This is the reading of most of the manuscripts. But some of the manuscripts and the 1633 print of Donne's *Poems* reads 'hands', so clearly there were those who thought that watches normally at this time had a minute hand as well.[8]

But apart from his familiarity with the mechanics of clockwork what is important about Donne's passage is that, for him, clocks and watches had important social functions: they exist to tell the populus how to 'employ their time'. What 'employ' means here is not entirely clear. It may refer to employment in the sense of the occupations of the townspeople. The regulation of the working day by means of public clocks had emerged in the fourteenth century.[9] Or 'employ' may mean the use of time in a more personal sense, a sense which saw idleness (the *accidia* of the medieval canonists) as a sin, or at least a fault. How one used one's time was a moral matter: as the author of *The Cloud of Unknowing* put it: 'take good keep unto tyme, how that thou dispendist it. For nothing is more precious than tyme. In oo litel tyme, as litil as it is, may heven be wonne and lost'. The verb 'dispendist' suggests accounting, and this meaning becomes clearer as the passage goes on. God will want to know, says the author, how one spent one's time 'in the dome and at the yevyng of acompte of dispendyng of tyme.'[10]

But most important of all is the paradox at the heart of this passage: though Donne knows that clocks and watches are devices used to regulate people's lives, he also knows that they are prone to being inaccurate and thus misleading. The pocket watch which goes wrong is compared to a sick person, who gets 'shaking palsies', whose sinews 'slacken', whose soul 'expires or languishes', whose pulse 'beats not or beats unevenly', whose voice 'rattles or grows dumb'. And the implication is that public clocks go wrong too. All this can be substantiated from contemporary records. In relation to watches Samuel Pepys is characteristically revealing. His reaction to acquiring a watch on 12 May 1665 is indicative of the fascination which clocks and watches held in the seventeenth century for those who wished to be fashionable. But important also are the misgivings he had about it:

> I cannot forbear carrying my watch in my hand in the coach all afternoon, and seeing what a-clock it is a 100 times. And am apt to think with myself: how could I be so long without one – though I remember since, I had one and found it a trouble, and resolved to carry one no more about me while I lived.[11]

7 See David S. Landes, *Revolution in time: clocks and the making of the modern world* (Cambridge, MA, 1983), pp 87–97 for the development of watches. For some illustrations of some watches with a single hand see F. J. Britten, *Old clocks and watches and their makers* (3rd ed. Woodbridge, 1994), pp 107–8. **8** For more on the development of minute hands, see Landes, pp 128–9. **9** On this see Jacques le Goff's classic essay 'Labour time in the "crisis" of the fourteenth century: from medieval time to modern time', in his *Time, work and culture in the Middle Ages*, trans. Arthur Goldhammer (Chicago, 1980), pp 43–52. **10** See *The Cloud of Unknowing and the Book of Privy Counselling*, ed. Phyllis Hodgson (EETS OS 218, 1944), p. 20. **11** See *Pepys's*

His misgivings were not misplaced. On 14 July 1665 he tells us that the watch was back at the clockmakers 'mending'. Good clocks and good clockmakers were rare at this time, and reliability was difficult to attain. As now, people tended to set their watches by reference to public clocks, but these among themselves tended to be inconsistent. An interesting reflex of this appears in Thomas Middleton's *Women Beware Women* (*c.*1621) as Bianca converses, in a desultory fashion, with her companions:

Bianca: How goes your watches, ladies; what's o'clock now?
First Lady: By mine, full nine.
Second Lady: By mine a quarter past.
First Lady: I set mine by St Mark's.
Second Lady: St Anthony's,
They say, goes truer.
First Lady: That's but your opinion, madam,
Because you love a gentleman o' th' name.
Second Lady: He's a true gentleman, then.
First Lady: So may he be
That comes to me tonight, for aught you know.
Bianca: I'll end this strife straight. I set mine by the sun;
I love to set by th' best, one shall not then
Be troubled to set often. (IV. i. 1–9)[12]

The passage has to be read carefully. There is more than a little sexual innuendo in it. It is also set in Florence, though the two churches mentioned are probably not Florentine – though they may be Italian.[13] But the basic issue of the unreliability of public clocks would have been instantly understood by a London audience, for London clocks were proverbially infamous for their inconsistency.[14] According to Bianca, the only reliable measure of time was the 'sun', that is time as calculated by the sundial, and it is no accident that Donne writes that had Harrington lived he might have proved to be a 'great sundial to have set us all', in terms of acting as a public example of virtue. At this time, clocks were regulated by sundials, which probably in part accounts for the large number of sundials which survive from the sixteenth and seventeenth centuries.

Diary, selected and ed. Robert Latham, 3 vols (London, 1996), II, pp 153, 166. **12** Thomas Middleton, *Women Beware Women*, in Bernard Beckerman (ed.), *Five plays of the English Renaissance* (1983; New York, 1993), pp 346–444. **13** J.R. Mulryne thinks that St Mark's in Venice and St Anthony's in Padua are probably behind these references, though St Mark's in Florence (opposite which the historical Bianca lived) is mentioned at I. iii. 84 in Middleton's play. There is no church of this period in Florence dedicated to St Anthony. **14** See Morris Palmer Tilley, *A dictionary of the proverbs in England in the sixteenth and seventeenth centuries* (Ann Arbor, MA, 1950), C426.

Yet, despite all this, despite the experience-based recognition that they were not infallible, that they frequently went wrong, mechanical clocks became associated with precision, reliability and perpetual motion – partly, no doubt, because they brought a new orderliness into timekeeping through using equinoctial hours, partly because they did not need as much attention as water-clocks or flame clocks, partly because if you could hear clocks striking the hours you had a constant reference point for the state of the day and the night. The mechanical clock, with its division of time into regular pulses so that every moment was the same as the last and the next, became a model for the logical arrangement of things, for unerroneous constancy, for temperance, for the properly ordered life. It gave writers an instrument which served as a religious, social and moral measure. Clocks could tell mankind how to organize its life in this world and prepare itself for the next. They came to be perceived as a threat too, an image of the remorselessness of the passing of things.

II

It is clear that engineers were working on the development of the mechanical clock in the second half of the thirteenth century: Robertus Anglicus in 1271 describes experiments being carried out to develop a mechanism driven by a falling weight.[15] Early clocks were based on this technology and were heavy and crude: the earliest clockmakers were non-specialist metalworkers – blacksmiths, locksmiths and gunmakers. The earliest known mechanical clock seems to have been that in the church of St Eustorgio in Milan in 1309.[16] But the instrument spread rapidly throughout western Europe, and specialist clockmakers emerged – John of Wallingford in England, Giovanni de' Dondi in Italy, Jean Fusoris in France. All of these developed intricate machines of great complexity, often involving more than simple horological movement: Giovanni de' Dondi's clock was an elaborate astrarium, as were a number of the earliest clocks – often involving programmes of religious, social and moral instruction. The astronomical clock in the town square in Prague, for example, which may date from as early as 1410, has a programme involving a number of illustrative features. In a lower roundel are shown the signs of the zodiac and the labours of the months, medieval programmes with long histories which suggest the unalterable nature of the annual cycle of the seasons and the appropriate, orderly and socially necessary tasks associated with them. But above, when the clock strikes, there is a marionette performance: a skeleton shakes an hourglass and rings a bell; a miser rattles his

15 See Jean Gimpel, *The medieval machine: the industrial revolution of the Middle Ages* (London, 1977), p. 153. **16** See Carlo Cipolla, *Clocks and culture, 1300–1700* (New York and London, 1978), pp 41–2.

moneybox; a vain man looks at himself in a mirror. Time and death, it seems to say, preside over the sinful world of mankind. Public clocks were not simply time-keepers; they sought to inculcate moral ways of living too. So also, on a smaller scale, did watches. Clocks and watches were perceived to figure the workings of the universe and sought to define man's place in it.

The earliest makers of astronomical clocks were implicitly comparing the universe to timepieces,[17] and it was not long before the comparison became explicit in literature. In his *Horologium Sapientiae* the German mystic Heinrich Suso (d. 1366) described a vision, which he dated as 1334, in which he saw Christ in the form of an elaborate contemporary clock chiming the hours: God's eternal time, he deduced, ought to provide a model for the soul which needed to act in consonance with it.[18] Earlier, in lines 10–18 of Canto XXIV of his *Paradiso*, Dante compared the circling of the Fellowship of the Lamb to the movement of the wheels of a clock which was evidently also an astrarium.[19] And in 1377, in his *Livre du Ciel et du Monde*, Nicole Oresme compared the universe to a clock (*horloge*) which worked whatever the season, by day or night, was never fast or slow, and which never stopped. And he pursued his own logic by making God into a clockmaker:

> Ainsi lessa Dieu les cielz estre meuz continuelment selon les proporcions que les vertus motivez ont aus resistences et selon l'ordenance etablie.[20]
>
> [In this manner did God allow the heavens to be moved continually according to the proportions of the motive powers to the resistances and according to the established order.]

The language here – 'motive forces', 'resistances' – demonstrates that Oresme is not using the comparison in an uninformed way, but that he understood the physics of clockwork.

Early clocks often consisted simply of a frame, which held the clockwork, without a case so that the workings of the machine could be seen, and there is evidence that watching the workings of a mechanical clock could be fascinating. Pondering the mechanisms of clocks in his *Treatise of Melancholie* (1591), for example, Timothy Bright is equally impressed by the paradoxical complexity and unity of them and sees a similarity with the workings of the universe:

17 See F.C. Haber, 'The cathedral clock and the cosmological metaphor', in J.T. Fraser and N. Lawrence (eds), *The study of time, II: proceedings of the second conference of the international society for the study of time, Lake Yamanaka, Japan* (Berlin, 1975), pp 377–416. **18** For a description of this treatise and the illustration which accompanies it see Eleanor P. Spencer, 'L'horloge de Sapience, Bruxelles, Bibliothèque Royale MS IV. iii', *Scriptorium*, 17 (1963), 277–99. See also H. Michel, '*L'horloge de Sapience et l'histoire de l'horlogerie*', *Physis*, 2 (1960), 291–8, where the clocks are discussed in great detail. **19** See *The Divine Comedy of Dante Alighieri*, with translation and comment by J.D. Sinclair, 3 vols (New York, 1981). On this passage see David Landes, *Revolution in time*, p. 57. **20** See Nicole Oresme, *Le livre du ciel et du monde*, ed. Albert D. Menut

> We see it evident in automaticall instrumentes, as clockes, watches, and larums, howe one right and straight motion, through the aptnesse of the firste wheele, not only causeth circular motion in the same, but in diverse others also: and not only so, but distinct in pace, and time of motion: some wheeles passing swifter than other some, by diverse rases; now to these devises, some other instrument added, as hammer, and bell, not only another right motion springeth thereof, as the stroke of the hammer, but sound also repeated, and delivered it at certaine times by equal pauses; and that either larume or houres according as the partes of the clocke are framed. To these if yet moreover a directorie hand be added; this first, and simple, and right motion by weight or straine, shall seeme not only to be author of deliberate sound, & to counterfet voyce, but also to point with the finger as much as it hath declared by sound. Besides these we see yet a third motion with reciprocation in the balance of the clocke. So many actions diverse in kinde rise from one simple first motion, by reason of variety of ioynts in one engine. If to these you adde what wit can devise, you may finde all the motion of heaven with the planets counterfetted, in a small modill, with distinction of time & season, as in the course of the heavenly bodies.[21]

Bright has clearly looked closely at the gearing of mechanical timekeepers and appreciates what can be driven by a weight or a spring ('straine') – a complex striking mechanism to register the hours or to act as an alarm, a hand on a dial, and an astrarium if desired, which might model 'the course of the heavenly bodies'. Implicit in this is the idea that the mechanical clock is like the universe, but though he makes that point in this passage here it is not his main point: a little later he returns to his point about unity and diversity and argues that the soul may seem to be three different souls, but is really one in which 'the same facultie varieth not by nature but by use only'.

So when Johannes Kepler asserted that 'the universe is not similar to a divine living being, but is similar to a clock' and when Robert Boyle described the universe as 'a great piece of clockwork' neither was using a particularly new analogy.[22] They had, however, more reason to use it than earlier writers, because by the seventeenth century the best clocks had improved somewhat in both accuracy and reliability.

and Alexander J. Denomy, trans. with an introduction by Albert D. Menut (Madison, WI, 1968), pp 288–9. I have used Menut's translation. **21** See *A treatise of melancholie*, with an introduction by Hardin Craig. The Facsimile Text Society No. 50 (New York, 1940), pp 67–8. I owe this reference to Dr Andrew Power. **22** For Kepler and Boyle see Carlo Cipolla, *Clocks and culture*, p. 105; see also *William Shakespeare: the sonnets and A Lover's Complaint*, ed. John Kerrigan (London, 1986), p. 36.

III

So the clock and later the watch became images for the logical workings of the universe, and the development of the mechanical method of calculating time also provided an incentive and an opportunity for seeking to establish greater political and social orderliness. In 1370 Charles V of France tried to organize time to his own standard: he decreed that all the clocks in Paris should be regulated by the one he was installing in the royal palace. The churches were to ring their bells when his clock struck the hour; the control of time by the church passed into secular hands. As Jacques le Goff put it, 'the new time became the time of the state'.[23] This situation was not helped, however, by the fact that the royal clock was not particularly accurate: *L'orloge du palais, / Elle va comme il lui plait.*[24] ['The clock of the palace goes as it pleases.'] So ran a contemporary satirical street-rhyme. To do what Charles V wanted to do was beyond both the technology of the age and its social organization, and the most interesting practical attempts to regulate society by regulating time concern nothing more complex than the hours of the working day.

Traditionally, the working day had been co-terminus with the hours of daylight, largely because of the unavailability, on a large scale, of reliable, efficient and inexpensive sources of artificial light. It was a tradition which grew up in relation to the rhythms of the natural world: it meant that the working day was longer in the summer than in the winter, which suited a largely agricultural economy though the tradition held in urban contexts too. Workers in Paris in 1395 were reminded, in no uncertain terms, that the working day was '*des heure de soleil levant jusqu'a heure de soleil couchant, et prenant leurs repas a heures raisonnables*' [from the hour of sunrise to the hour of sunset, with meals taken at reasonable times]. But work-clocks had been installed in many towns in northern Europe in the fourteenth century – in Arras in 1315, in Ghent in 1324, in Amiens and Aix-sur-Lys in 1335, in York between 1352 and 1370. The clocks became the site for labour disputes: fines were levied on workers who did not obey the summons of the clock in some towns. But there was resistance too: on 16 March 1367, in order to head off a strike, the dean and chapter of Thérouanne had to promise the 'workers, fullers, and other mechanics' that they would 'silence forever the workers' bell'.[25] Theoretically, the use of the mechanical clock as a social regulator ought to have made it possible to negotiate hours of work which did not correspond to the hours of daylight or to what were called 'working days', yet those who chose to work at night, or on the Sabbath, or on 'holy days' still came under

23 See Jacques le Goff, *Time, work and culture in the Middle Ages*, trans. Arthur Goldhammer (Chicago, 1980), pp 49–50. **24** Quoted by Carlo Cipolla, *Clocks and culture*, p. 41, and David Landes, *Revolution in time*, p. 75. **25** For an excellent account of these labour regulations and disputes see Jacques le Goff, *Time, work and culture in the Middle Ages*, pp 44–9.

some social and moral stigma. In an early fifteenth-century alliterative poem the author sees as diabolical the work which goes on in a blacksmith's shop operating at night: 'the devel it to dryve'.[26]

From much later comes a highly imaginative, and famous, confrontation which turns on the issue of when it is appropriate to work, a confrontation which specifically invokes the mechanical clock and is largely defined by it. In Shakespeare's *I Henry IV*, when Prince Hal and Sir John Falstaff first appear on stage together, they talk about time, and its licit and illicit uses:

> *Sir John*: Now, Hal, what time of day is it, lad?
>
> *Prince Harry*: Thou art so fat-witted with drinking of old sack, and unbuttoning thee after supper, and sleeping on benches after noon, that thou hast forgotten to demand that truly which thou wouldst truly know. What a devil hast thou to do with the time of day? Unless hours were cups of sack, and minutes capons, and clocks the tongues of bawds, and dials the signs of leaping houses, and the blessed sun himself a fair hot wench in flame-coloured taffeta, I see no reason why thou shouldst be so superfluous to demand the time of day. (II. ii. 1–12)[27]

Hal responds to Falstaff's question about 'the time of day' by anatomizing his daytime activities in terms of a mechanical timekeeper, which had both a bell ('clock') and a 'dial': 'sleeping upon benches after noon' drinking 'sack', eating 'capons', talking to 'bawds' and going to brothels ('leaping houses'). Falstaff does no work during the day, but merely indulges himself when he is awake. The clock was obviously assumed to have a hand which marked the hours of daylight, often on a separate dial, which had on it 'the blessed sun himself', and it is this feature which determines Falstaff's reply and defines the shape of the scene as it develops. Falstaff admits, tacitly, that his lifestyle does not involve spending the daylight hours working, but says that he works at night, and alludes to the fact that many contemporary clocks had a dial for the hours of night, which were indicated by a hand with a moon on it: 'Indeed you come near me now, Hal, for we that take purses go by the moon and the seven stars', he says, 'and not "By Phoebus, he, that wand'ring knight so fair."' (I. ii. 13–15) Where the iambic pentameter line about Phoebus quoted by Falstaff comes from is not known, but it is clear that he is rejecting the daytime world of work for the night-time world of crime. And he follows this by seeking, only half humorously, to persuade Hal, when he is king, to give sanction to such activities as he indulges in by inventing parodic titles of

26 For the poem see R.H. Robbins (ed.), *Secular lyrics of the fourteenth and fifteenth centuries* (2nd ed. Oxford, 1964), pp 106–7. **27** See Stanley Wells and Gary Taylor (eds), *The Oxford Shakespeare: the complete works* (1988; Oxford, 1998), p. 456.

actual royal offices – 'squires of the night's body', 'Diana's foresters', 'gentlemen of the shade'. But Hal reminds him that crime leads not to titles, but to prison or the gallows. Falstaff defends 'purse-taking' as his vocation: ''Tis no sin for a man to labour in his vocation' (105).

But Hal is not to be deflected, and in the famous closing soliloquy to this scene (192–214) he commits himself to work and to day-time values associated with the sun, and determines that he will begin 'redeeming time' which he knows he has hitherto wasted. He will put up with the 'unyoked' (i.e. not working) 'idleness' of his companions, but for 'a while' only. If 'playing holidays' become everyday things, he argues, they become 'tedious'. He vows instead to 'imitate the sun'. In this context such a statement has multiple resonances. One, by way of a pun, is that he will become the dutiful and responsible 'son' that his father wishes him to be. Another is that he will live up to his family traditions, since the 'sunburst', or the sun emerging from clouds, was a Plantagenet badge. But in view of the way the clock has shaped this scene and in view of the vocabulary of the final soliloquy, the commitment to work of a licit and useful kind is important in his scale of priorities: he intends to conform, to commit himself to order and regularity, to become a useful member of the community, to live his life in relation to the instrument which had come to govern social life. Falstaff has chosen a different way, and as the plays develop there are reminders of this choice. In the play-acting scene in Eastcheap, Falstaff, in the role of Hal's father marvels 'where thou spendest thy time' and accuses Hal of truancy using this imagery: 'Shall the blessed sun of heaven prove a micher, and eat blackberries?' (II. v. 411–12). In Falstaff's painful interview with the Lord Chief Justice, the latter alludes both to Falstaff's part in the Gads Hill robbery and his (supposed) good performance at the battle of Shrewsbury in the following terms: 'Your day's service at Shrewsbury hath a little gilded over you night's exploit at Gads Hill' (*II Henry IV*, I. ii. 149). In fact, throughout *II Henry IV*, whenever Hal is with his old companions, he is haunted by the feeling that he is not spending his time profitably. In the course of a desultory conversation with Poins emerging from a discussion of a letter from Falstaff, he pulls himself up short with: 'Well, thus we play the fools with time and the spirits of the wise sit in the clouds and mock us' (II. ii. 133–4). And again, a little later in Eastcheap, he confesses to Poins that he feels 'much to blame / So idly to profane the precious time' (II. iv. 364–5), when there is national business he should be attending to at court.

IV

How one spent one's time was a moral as well as a social matter, then, and there are numerous literary examples from the Middle Ages of clocks being used as symbols of temperance and constancy. As models for moral behaviour, however, timekeepers are best known in the sixteenth and seventeenth centuries because of the use emblem writers made of them. Time was one of the most popular subjects in emblem books, and time as a personified concept appeared in many guises – as an old man with a scythe ('Father Time'), as a winged figure ('Time flies'), or as a figure associated with instruments for measuring time. The traditional instruments appear – sundials, sandglasses (sometimes with wings), clepsydrae – mostly in relation to the swift passage of time, the course of human life and the ages of man, to the proper and improper uses of time. And, unsurprisingly, mechanical clocks begin to be used in the same contexts.

Many examples could be cited but one must serve for many. In an example from Sebastian de Covarrubias Orozco's *Emblemas Morales* (1610), it is the physics of a clock's movement, principally the weights, which generates the analogy. The poem is in praise of *gravitas* in various punning senses of the word:

> Anda el relox de pesas mas ligero,
> Quando ellas son mas graves y pesadas,
> El hombre quanto mas grave y entero,
> Tanto mas assegura sus pisadas,
> Agil, firme, constante, y verdadero,
> Senalando sus horas compassadas,
> Enfin es un relox, tan regulado,
> Que tarde, o nunca esta desconcertado.[28]

> [The clock with weights goes more quickly when the weights are heavier and more substantial. The man, insofar as he is heavier and more robust, is so much more secure in his footsteps, nimble, stable, constant, and true, marking out his measured hours. He is, just like a clock, so regulated that he is never slow or disordered.]

The well-regulated clock is a model for a well-ordered life: the perfection of the machine is something for a human to aspire to and emulate if possible. It is hard to imagine a more complete triumph for the mechanistic view of the world than this little poem.

But for those who did not spend time appropriately, or those who wasted time, the clock became a potentially menacing image. In Shakespeare's *Richard II*, as Richard languishes in prison in Pomfret Castle after having been forced to give up his throne, the deposed king meditates on the subject:

28 For the emblem and the verse see Arthur Henkel and Albrecht Schloe (eds), *Emblemata: handbuch zur sinnbildkunst des XVI. und XVII. jahrhunderts* (Stuttgart, 1976), col. 1340. The translation is mine.

I wasted time, and now doth time waste me,
For now hath time made me his numb'ring clock.
My thoughts are minutes, and with sighs they jar
Their watches on unto mine eyes, the outward watch
Whereto my finger, like a dial's point,
Is pointing still in cleansing them from tears.
Now, sir, the sounds that tell what hour it is
Are clamorous groans that strike upon my heart,
Which is the bell. So sighs, and tears, and groans
Show minutes, hours, and times. But my time
Runs posting on in Bolingbroke's proud joy,
While I stand fooling here, his jack of the clock. (V. v. 49–60)

To make the man who is hyper-conscious of the passage of time and the way time is his enemy into a clock is bold and imaginative: the passage of time and Richard II's misery, as the allegory unfolds, become co-terminous: 'sighs, and tears, and groans / Show minutes, hours and times' in a one-to-one equivalence. But the passage has resonances beyond the central comparison. The transposed grammatical equivalences of the line 'I wasted time, and now doth time waste me' suggests that Richard may feel that what has happened to him is part of an unforgiving 'eye for an eye' vengeance. It may also allude to the stories that Richard II was starved to death ('waste') in Pomfret Castle. But the deposed king is acutely conscious, also, of his lost status as the controller of time, and one recalls his commutation of four years of Bolingbroke's original banishment and Bolingbroke's sardonic comment on it:

How long a time lies in one little word!
Four lagging winters and four wanton springs
End in a word: such is the breath of kings. (I. iii. 206–8)

But Richard II's time is now controlled by Bolingbroke. When he says it 'runs on in Bolingbroke's proud joy' he may intend no more than that while he is imprisoned Bolingbroke is going forever to enjoy kingship. But Shakespeare may also be alluding to the way in which kings had sometimes, following Charles V in 1370, tried to standardize time, or to the way in which noblemen and other dignitaries often had themselves painted with small table-clocks in the background, as status symbols and expensive toys.[29] Richard II clearly feels that in some way he now belongs to Bolingbroke, and by making himself not only into a clock but

29 See, for example, Titian's 'Knight of Malta' in the Prado, Madrid. For a reproduction see Carlo Cipolla, *Clocks and culture*, p. 80, and for details of other portraits with clocks see p. 49.

into a 'jack of the clock' he is enhancing this impression. Clockjacks were usually dressed as servingmen or labourers, manual workers doing a heavy job with hammers, and their function in relation to public clocks was, in part at least, to amuse bystanders.[30] The comparison with a clock suggests, not only that the deposed king is conscious, in retrospect, of his wasting of time, but that he feels that he may have become a symbol of Bolingbroke's newly acquired power and a laughing stock, 'fooling' for the amusement of the new king and others.

V

The clock enabled people to measure time more accurately, but it did not enable them to control it any the better. Quite the reverse in fact. The regular and incessant ticking of the clock, measuring the time in exact pulses without reference to anything except its own humanly constructed mechanism, brought a new sense of anxiety into ideas about the passage of time: the clock not only measured out the progress of a human life but, minute by minute, ticked it on to death. The clockmakers had invented not only a useful instrument, but one that was also, at best, disconcerting and, at worst, terrifying.

And this was recognized by clockmakers themselves. The Prague clock's striking mechanism is activated by a marionette of a skeleton which pulls a rope. But personalized timekeepers, precisely because they are personalized, more often make the point, usually by way of engravings. A table watch given by Sir William Cooper to Elizabeth Cooper, his daughter-in-law, in 1539 has a coat of arms on the cover and engravings on the dial. The figure of Christ is accompanied by emblems of death and two mottoes: '*Vigilate et orate quia nescitis horam*' ['Watch and pray for you know not when the time is', Luke xiii. 33] and '*Quae libet hora ad mortem vestigium*' ['Every hour is a step towards death'].[31] *Memento mori* watches in the shape of skulls became fairly common in the sixteenth and seventeenth centuries: people carried reminders of death in their pockets. One such, made by Pierre Moysant (*alias* Moyse) of Blois, is very elaborate.[32] The skull is silver gilt and is generously incised. On the skull's forehead is Death as a skeleton, with his scythe and sandglass, standing between a cottage and a palace, and around the picture is the following text from Horace: '*Pallida mors aequo pulsat pede pauperum tabernas regumque turres.*' ['Pale death visits with impartial foot the cottages of the poor and the towers of the rich', *Odes* I. iv. 13–14] On the back of the skull is Time, again with a scythe, and another text from Horace: '*Tempus*

30 For an interesting account, with illustrations, of mainly English clockjacks see F.J. Britten, *Old clocks and watches and their makers*, pp 36–58. **31** For an account of this clock see F.J. Britten, *Old clocks and watches and their makers*, p. 103. **32** For the details of this macabre object, and illustrations, see Britten, *Old clocks and watches and their makers*, pp 109–11. For another illustration see Landes, *Revolution in time*, Fig. 11.

edax rerum tuque invidiosa vetustas.' ['Time and you, envious old age, are devourers of all things', *Metamorphoses*, XV. 234] There is a serpent with its tail in its mouth, an emblem of eternity, incised on this panel too. In the two panels on the top of the skull appear the fall of man, which brought death into the world, and the crucifixion, with an inscription stressing that death has been overcome: '*Sic justiciae satis fecit, mortem superavit, salutem comparavit.*' ['Thus justice was satisfied, death overcome, and salvation obtained.'] Emblems of the crucifixion run round the openwork beneath the four panels. Inside the watchcase is the holy family with Jesus in the manger, and on the silver dial appears Saturn eating his children – time the devourer.

It can be argued that all clocks stress the inexorable passage of time and the inevitability of death, but watches of this sort make it especially plain. And it is probably a watch of this sort that Ciro di Pers is thinking of when he writes:

> Nobile ordigno di dentate rote
> Lacera il giorno e lo divide in ore,
> Ed ha scritto di fuor con fosche note
> A chi legger le sa: SEMPRE SI MORE. (1–4)[33]

> [The excellent instrument with toothed cogs tears the day and divides it into hours, and it has written on the outside with black letters, for whoever can read them, WE ARE DYING ALWAYS.]

The motto on the watch case is perhaps best understood as the continuous present tense, because this is very much a poem not about sudden death but about the inexorable step-by-step journey to death through time, marked out by the movement of the clock: the idea of time the devourer is present here in a very precise way in relation to the toothed (*dentate*) cogs of the clock's movement which chew up the day, and this idea returns later in the sonnet in the phrase *eta vorace* ['voracious age']. But it is the ceaseless sound of the clock which grips his attention, the ticking in the 'hollow metal' (*metallo concavo*) is like a 'funereal voice' (*voce funesta*) which 'echoes' (*risuona*) in his heart. They are like instruments of war, a drum or a trumpet, daring him onwards, and it is with metallic sound that the poem closes:

> E con que'colpi onde 'l metal rimbomba,
> Affretta il corso al secola fugace,
> E perche s'apra, ognor piccia alla tomba. (12–14)

> [And with those beats from which the metal resounds, one's progress hastens through flying time, and each hour taps on the tomb so that it will open.]

33 See Benedetto Croce (ed.), *Lirici Marinisti* (Bari, 1910), p. 372.

What desolates the writer of this poem is the sheer mechanical repetitiousness of the time the clock delivers, of which one is reminded in every tick.

How it appeared to the original audience is something that one can only speculate about, but the striking of the mechanical clock in the last scene of Christopher Marlowe's *Dr Faustus* retains its ability to shock at once by its surprise and by the inevitability of its logic. The coming of death is a traditional feature of morality drama, but what gave Marlowe the idea of using the clock can only be guessed at. Perhaps because the story is set in Germany and that German clocks were imported into England in the latter part of the sixteenth century suggested it. Or perhaps the fact that Faustus is a scholar and that the play begins and ends with him in his study had something to do with it, for clocks and other instruments for measuring time were often found in renaissance studies and workrooms: a fourteenth-century Florentine humanist had said that no study should be without a clock. Or perhaps the twenty-four years' contract with the devil, which Marlowe inherited from the traditional story, suggested an analogy with the hours of the day. In a sense, Faustus has had control over time for twenty-four years: it has enabled him to make present figures from the past (Alexander and his concubine, XII. 54; Helen of Troy, XVIII. 28, 99), and grapes out of season (XVII. 21).[34] But he has scarcely done anything important with this ability. Thomas à Kempis had written, reproving a friend: 'It is sad that you do not employ your time better, when you may win eternal life hereafter. The time will come when you will long for one day or one hour in which to mend, and who knows whether it will be granted.'[35] It is this process of wishing for more time, too late, that Faustus goes through on his last night as the clock strikes eleven:

> Now hast thou but one bare hour to live
> And then thou must be damn'd perpetually.
> Stand still, you ever-moving spheres of heaven,
> That time may cease, and midnight never come;
> Fair nature's eye, rise, rise, again and make
> Perpetual day; or let this hour be but
> A year, a month, a week, a natural day,
> That Faustus may repent and save his soul. (XIX. 134–41)

He quotes a version of a line from Ovid's *Amores*, I. xii. 40: '*O lente, lente currite noctis equi*' ['O run slowly, slowly, horses of the night!'] where Ovid tries to prolong a pleasurable night he has spent with his lover – unavailingly. It could be

34 The references here are to *Dr Faustus*, ed. John D. Jump (London, 1962). 35 Quoted in Landes, *Revolution in time*, pp 90–1.

done – but only if you were a god: in *Amores* I. xiii. 45–6, Jove artificially extended the night he spent with Alcmena, when Hercules was conceived, by ordering the chariot of the moon make three circuits of the heavens. Faustus had thought that his pact with the devil and his abilities at the magical arts might make him at least into a demigod: 'A sound magician is a demi-god; / Here tire, my brains, to get a deity!' (I. 61–2) But Faustus is neither a deity nor a demigod. Nor does he exist in the mythological world of Ovidian time. His time is organized by the striking of the clock, the contemporary mechanical timekeeper which had its own scientific momentum. The time of the clock is not reversible, or extendable, or stoppable: 'The stars move still, time runs, the clock will strike.' (XIX. 143). No evasion is possible for Faustus, no escape, and the clock strikes, in turn, the half hour and finally midnight: 'O, it strikes, it strikes! Now, body, turn to air, / Or Lucifer will bear thee quick to hell!' (183–4). The transaction in which Faustus has engaged with the devil is formulated rather like a business agreement. As he explains to the scholars earlier in the scene: 'I writ him a bill with mine own blood: the date is expired; the time will come and he will fetch me' (66–8). So it is perhaps appropriate that the mechanical clock, which had come to regulate business, should be the instrument by which Faustus is held to account, but it is clear that Marlowe had been thinking seriously about clocks and closure, perhaps on a more metaphysical or philosophical level, from the laconic signing off formula for the play: '*Terminat hora diem, terminat auctor opus*' ['The hour ends the day, the author ends his work'] – though this again is rendered in terms of the worker's being released from toil at a predetermined moment, timed by the clock.

'Shakespeare conceded, more readily than Spenser or any of his great contemporaries, the mechanical origins of his anxiety', writes John Kerrigan, in relation to the appearance and importance of the clock in the *Sonnets*.[36] In fact, a whole range of systems relating to the passing of time is invoked – the turning of the seasons, the labours of the months, the lunar cycle, the course of the sun through the day, the ages of man, and other sorts of timekeepers are referred to, particularly the sundial and the sandglass. But, especially in the sequence of sonnets 1–126, the clock is a powerful presence, as Shakespeare tries to persuade the 'fair young man' to recognize the aging process, to marry and have children as a way, along with others, of overcoming time and achieving for himself some sort of immortality. When the clock appears it is, significantly enough, in Sonnet 12, and its opening line stresses counting, the striking of a clock which registers the twenty-four-hour cycle in two sets of twelve: 'When I do count the clock that tells the time' (1). The mechanical clock appears again, significantly in Sonnet 60, which deals with 'minutes', and many of the images and themes of the earlier

36 See *William Shakespeare: The sonnets and A Lover's Complaint*, ed. John Kerrigan, p. 38. I am much indebted to John Kerrigan's excellent account of time in the *Sonnets* on pp 33–41.

sonnet reappear: 'Like as the waves make towards the pebbled shore, / So do our minutes hasten to their end' (1–2). The first clocks with minute hands appeared in the sixteenth century and it is one of these which Shakespeare has in mind, the contemplation of which, with the number of the sonnet, generates the opening.[37] Time is here broken down into ever smaller segments.

None of the other sonnets begins with the clock, but the clock, nevertheless, haunts the sequence. In Sonnet 115, for example, 'reckoning Time', that is, time that counts and is counted, and also time that holds one to account in a legal and mercantile sense, appears and it is a destructive figure:

> [...] reckoning Time, whose millioned accidents
> Creep in 'twixt vows and change decrees of kings,
> Tan sacred beauty, blunt the sharp'st intents,
> Divert strong minds to th'course of alt'ring things – (5–7)

The word 'creep' alludes to the saying 'Time creeps', and 'tan' suggests the aging process. All this appears again, in a slightly altered form, in the sestet of Sonnet 104, another text about aging and the passage of time:

> [...] yet doth beauty, like a dial hand,
> Steal from his figure, and no pace perceived;
> So your sweet hue, which methinks still doth stand,
> Hath motion, and mine eye may be deceived;
> For fear of which, hear this, thou age unbred:
> Ere you were born was beauty's summer dead. (9–14)

The idea of seasonal time is obviously present in the last phrase, but much of this relates to timekeeping. The 'dial hand' is the hand of the mechanical clock, which 'steals', in the proverbial sense,[38] because it is difficult to perceive its slow but unstoppable movement ('no pace perceived'), just as it is difficult to perceive, in the short term, the aging process: it may appear that the young man's beauty 'still doth stand' but that is an illusion – the poet may be 'deceived' – because actually the process of aging 'hath motion', and continues whether one can register it in the short term or not. It is the constant ticking away of time which cannot be reversed which proved to be the most terrifying aspect of the mechanical clock.

This theme may even appear, in an aetiolated form, in the last poem of this sequence, Sonnet 126, which pulls together many of the ideas and images about time which run through the whole work[39]:

37 See René Gratiani, 'The numbering of Shakespeare's sonnets: 12, 60, and 126', *Shakespeare Quarterly* 35 (1984), 79–82. 38 Compare B.J. and H.W. Whiting, *Proverbs, sentences and proverbial phrases from English writings mainly before 1500* (Cambridge, MA 1968), T325. 39 See Jon R. Russ, 'Time's attributes in

O thou, my lovely boy, who in thy power
Dost hold Time's fickle glass, his sickle hour,
Who hast by waning grown, and therein show'st
Thy loveers withering, as thy sweet self grow'st;
If Nature, soverign mistress over wrack,
As thou goest onwards, still will pluck thee back,
She keeps thee to this purpose, that her skill
May Time disgrace and wretched minutes kill.
Yet fear her, O thou minion of her pleasure;
She may detain, but not still keep her treasure.
Her audit, though delayed, answered must be,
And her quietus is to render thee.

By a bold metaphorical transference Shakespeare makes the 'lovely boy' into a figure of time: he holds the hourglass and the sickle. He has grown older, more impressive, as time has passed and as his lovers have been 'withering' – ostensibly a seasonal metaphor which is picked up later. But 'sickle' and 'waning' also suggest movement through time by way of the lunar cycle, and 'waning' and growing here have been read off at one level as referring back to the hourglass: as the top compartment gets emptier, the pile of material in the bottom compartment grows. The clock is suggested by 'minutes', and the argument, briefly, seems to allow that 'Nature', who keeps the boy looking young may be intending in this case to 'kill' time, reversing the normal processes. But the last four lines recognize the inevitability of the process of aging: Nature can 'detain', in the sense of 'slow down', but not keep him 'still', that is, 'in a static state' as well as 'constantly'. In the last couplet 'audit' and 'quietus' are accounting words and the idea of paying one's debt to nature, that is, 'dying' is present by implication. Though it concludes a sequence of sonnets, this poem is not in the conventional fourteen-line form: it consists of twelve lines in couplets. William Thorpe, in 1609, printed it with two sets of brackets where lines 13 and 14 might have been, as if something was missing. But the poem makes complete sense as it is, and it may well be, since he had used the number 12 earlier to generate a poem on a clock, that Shakespeare wanted this numerical association in the final poem of this sequence.

As clockmakers' skills developed timepieces became increasingly accurate, and it became possible to divide time into ever smaller units: clocks with second hands became increasingly available in the seventeenth century. And it is the small divisions of time which generate the anxiety behind Lord Herbert of Cherbury's poem 'To his Watch when he Could not Sleep':

Shakespeare's sonnet 126', *English Studies* 52 (1971), 318–23.

Uncessant minutes, whil'st you move you tell
The time that tells our life, which though it run
Never so fast of farr, your new begun
Short steps shall overtake; for though life well

May scape his own account, it shall not yours,
You are Death's auditors, that both divide
And summ what ere that life inspir'd endures
Past a beginning, and through you we bide

The doom of Fate, whose unrecall'd decree
You date, bring, execute; making what's new,
Ill and good, old, for as we die in you
You die in Time, Time in Eternity.[40]

This is again about counting and accounting: 'tell' again here means 'communicate' as well as 'count out', and the idea of death as the closing of a financial account is again present throughout. But it is the 'uncessant' nature of mechanical time, the 'minutes' steadily ticking by which seizes Lord Herbert's attention. They are 'Death's auditors' because they both 'divide' and add up ('summ') the course of a life. And they operate without any moral distinction, counting out whatever is 'ill and good' in the same mechanical way. The end of the poem may be intended to be consolatory: 'minutes' disappear in the more general flux of time, and time itself may be subsumed in eternity, but it does not cancel the disturbing paradox that the 'short steps' of time have the ability to 'overtake' life 'though it run / Never so fast or farr'. Mechanical time had its own momentum and nothing demonstrated this more insistently than the personalized timekeeper which became an ever-present accoutrement for the better off: when it was not in their pockets it was on their bedside tables.

40 See Helen Gardner (ed.), *The metaphysical poets* (London, 1957), p. 92.

Bodies of knowledge and knowledge of the body in sixteenth-century literature

AMANDA PIESSE

> What piece of work is a man, how noble in reason, how infinite in faculties, in form and moving how express and admirable, in action how like an angel, in apprehension how like a god: the beauty of the world, the paragon of animals – and yet to me what is this quintessence of dust?
>
> William Shakespeare, *Hamlet* (II. ii. 303–8)[1]

Hamlet's speech is characteristically qualified; after celebrating and marvelling at everything that humankind is, we are brought straight back to the prayer-book remembrance that humankind is made of dust, and that to dust it will return. In another characteristically paradoxical twist, we are reminded that it is something unknown, perhaps unknowable, probably mystic, the 'quintessence' that animates this earthly, earthy body. By 1603, when the play was first printed, there was a real tension between what was known about the human body (it is now some sixty years after the publication of Andreas Vesalius' *De humani corporis fabrica* [1543], which lays bare in extraordinary detail both the skeletal and the muscular workings of the human frame, as we shall see), and how that knowledge sits alongside what is not known about the nature of humankind. We see the tension in the scientific accounts, too; it is set out as clearly in the works of medics as in the works of playwrights and poets. Both John Banister, who was a surgeon, and Caspar Barlaeus, an anatomist, realized that though the physical frame is known, what humankind is still remains a mystery.

Banister, in his *Historie of Man* (1578), describes the project of his text as follows: 'I have earnestly, though rudely, endeavoured to set wide open the closet door of nature's secrets, whereinto every godly artist may safely enter, to see clearly all the parts, and notable devices of nature in the body of man'.[2] Barlaeus, writing in 1638, has the following to say about the art of anatomizing:

1 All references to the play, unless indicated otherwise, are to William Shakespeare, *Hamlet* (1982; London, 1997). 2 John Banister, *The historie of man, sucked from the sappe of the most aproued anathomistes* (London,

> Here, keenly intent, we behold whatever we are inside; and that which is hidden in our bodies' fabrica is brought to light. This structure is the seat of the soul, this is the venerable tabernacle of the mind, this insignificant receptacle contains divinity concealed. Whatever we were, we are no longer.[3]

Hamlet, on the subject of the body as on just about every other matter for debate in the play, goes back again and again to the unknowability of the human condition. In a fierce exchange with Guildenstern, he argues bitterly for this unknowability:

> *Hamlet*: … Will you play upon this pipe?
> *Guildenstern*: My lord, I cannot.
> *Hamlet*: I pray you.
> *Guildenstern*: Believe me, I cannot.
> *Hamlet*: I do beseech you.
> *Guildenstern*: I know no touch of it, my lord.
> *Hamlet*: It is as easy as lying. Govern these ventages with your fingers and thumb, give it breath with your mouth, and it will discourse most eloquent music. Look you, these are the stops.
> *Guildenstern*: But these cannot I command to any utterance of harmony. I have not the skill.
> *Hamlet*: Why, look you now, how unworthy a thing you make of me. You would play upon me, you would seem to know my stops, you would pluck out the heart of my mystery, you would sound me from my lowest note to the top of my compass; and there is much music, excellent voice in this little organ, yet cannot you make it speak. 'Sblood, do you think I am easier to be played on than a pipe? (III. ii. 340–361)

The question here is this; what is it to know the nature of humanity, or of an individual person? Is it to know the body, or to know the mind, or to be able to place them in a particular context? Which is the greater question, to know what humankind is, or to know what a person is?

Hamlet – both the person and the play – is notoriously ambiguous in nature, and the shifting ground that Shakespeare creates for this particular examination of human nature is perhaps best summed up, for our purposes, in one of Hamlet's letters to Ophelia, read here by Polonius:

1578) sig. BI verso. Cit. Michael Neill, *Issues of death: mortality and identity in English renaissance tragedy* (Oxford, 1998), p. 123. 3 Caspar Barlaeus, 'On the anatomical house which can be visited in Amsterdam' 1638, Fig. II, quoted in William S. Heckscher, *Rembrandt's anatomy of Dr Nicolaas Tulp* (New York, 1958), p. 114 and Michael Neill, p. 102.

'Doubt thou the stars are fire,
Doubt that the sun doth move,
Doubt truth to be a liar,
But never doubt I love.
O dear Ophelia, I am ill at these numbers. I have not art to reckon my groans. But that I love thee best, O most best, believe it. Adieu.
Thine evermore, most dear lady, whilst this
machine is to him, Hamlet.' (II. i. 115–123)

This is the beginning of the seventeenth century. Since the publication of Vesalius' *De humani corporis fabrica* and Copernicus' *De revolutionibus orbium coelestium*, both in 1543, it has been clear that the body is, essentially, a machine; but we are also clear, even if it is perhaps not wise to say so publicly, that there is plenty of evidence to suggest that the sun does not orbit the earth. Shakespeare holds two periods in time in tension; the time that the play represents, the tenth century, and the time in which it is being written, the early seventeenth century. So perhaps we are momentarily being presented with questions about the nature of time here too – what is the present of this play, the time it reflects, the time in which it is written, or the time at which we observe it? A twelfth-century person surely *wouldn't* doubt that the sun doth move; but an early-seventeenth-century one would be well behind the times if she was not at least wondering about it. Or perhaps the point is that popular belief does not necessarily move in tandem with knowledge. Or that, like Hamlet, we are 'ill at these numbers': not able to make the mathematical computations necessary, unable to write good poetry (it means both, at the same time), sick of trying to make sense of things either by science or by art, or caught between the two, just as we are, as human beings, both feeling, thinking, confused, inexplicably complicated beings, and at the same time, completely understandable, complicated but explicable and demonstrable machines.

What is clear to me from this is that the scientific investigation of the body and the more metaphysical investigation of what it is to be 'human' never really separates out fully during the sixteenth century, and this is as true of scientific works as it is of literary works. In what follows, therefore, I want first to consider the taking apart of the physical body through the process of dissection, and all the cultural baggage that attaches to that, by looking at the work of Andreas Vesalius.[4] Then I want to think about how sixteenth-century humankind has to realign its notion of itself when it realizes that, thanks to Copernicus, Galileo and Kepler, the earth can no longer be said to be at the centre of the universe. Finally, I want

4 The website http://vesalius.northwestern.edu/noflash.html is the easiest and most comprehensive way to access the text of *De humani corporis fabrica*. References to the text cited here are to this internet source, accessed 22 August 2006.

to see how plays and poems of the period respond to and reiterate these challenges to the idea of what it is to be human.

* * *

During the sixteenth century, three or four shifts in scientific and technological knowledge had a huge impact on the way in which humankind considers itself. The calculations of Copernicus, as published in *De revolutionibus orbium coelestis* and the observations of Galileo as set out in *De motu* in about 1590 displaced humankind from its position at the centre of a fixed universe. Galileo's later observations were only possible because of advances in glassmaking and lens-grinding, as Copernicus' were prompted by a renewed interest in the Ptolemaic system reawakened by Regiomontanus' *Epitome of the Almagest*, a work which brings together literary humanist concerns with those of astronomy and mathematics, and one which examines 'the glories of antiquity and the contrasting cultural poverty of the [Renaissance] present'.[5] Not unconnected from the work of both men is the reconsideration and realignment of Western thought in a new paradigm of relative knowledge. The fall of Constantinople in 1453 brought about rediscovered terms of reference, and the possibility of returning to original texts in Greek rather than their corrupt Latin translations.[6] Movements forward in mapmaking coupled with a new accuracy in nautical instruments led to voyages of exploration that in turn caused Western thinkers to reconsider their position in relation to competing civilizations in hitherto unexplored parts of the world.[7] As the invention and proliferation of moveable type allowed for a greater dissemination of texts, so the variety of texts available for consumption expanded, and with such availability came greater consumption and greater, more varied demand.[8] As the boundaries and understanding of the exterior world increased, so the very same prompts caused a journey inwards too. By the end of the sixteenth century, improved lens techniques allowed for investigation with rudimentary microscopes, which, while so crude as to be little more than toys for the next hundred years or so, still held up the possibility of the inward as well as the outward journey by means of what was essentially the same technique. In terms of paradigms of knowledge, shifts in the methodology of thinking were brought about by increased pressure to be empirical, caused by renewed interest in the relationship

5 Peter Dear, *Revolutionising the sciences: European knowledge and its ambitions 1500–1700* (Basingstoke, 2001), p. 33. **6** For a brief and convincing account of the reawakening of interest in original Greek texts and its effect on the Aristotelian traditions in natural philosophy, see Peter Dear, pp 44–5, and Eugene F. Rice and Anthony Grafton, in *The foundations of early modern Europe, 1460–1559*, 2nd ed. (New York and London, 1994), pp 22–5. Dear concludes his argument (p. 45): 'In keeping with Melanchthon's humanist predilections, Aristotle began to be studied with much closer concern for fidelity to the original Greek and to related philological matters. The original Aristotle was just as important to recover as the original Ptolemy – or the uncorrupted text of the Bible.' **7** See Rice and Grafton, pp 26–32. **8** See Rice and Grafton, p. 24.

between the practical and the theoretical because of both renewed availability of philosophical work and a new kind of writing as print technology put the possibility of textual production in hands other than those of the church and the universities. Combined, these events saw the advent and advancement of the technician-philosopher as a new kind of thinker.[9] In the words of Rice and Grafton:

> Artists-engineers won a higher social position than they had enjoyed before. Their artistic and technological achievements aroused the interests of both literary men and scholastic philosophers and diminished the traditional contempt of the academically learned for manual and mechanical operations.[10]

One place in particular where all these shifts meet is in the work of anatomy.

As Jonathan Sawday points out, the anatomist brings about the birth of a science which 'transforms entirely people's understanding not only of themselves but of the relationship of their minds to their bodies and even their feelings of location in human society and the natural world.' Anatomization is a paradox; 'as the physical body is fragmented, so the body of understanding is shaped and formed' while in the anatomy theatre the human body is replaced as the central object of the little universe, just as it is being displaced by Copernican theory.[11] Vesalius' frontispiece to his 1543 edition of *De humani corporis fabrica* shows a public dissection taking place, the corpse displayed as if on stage, presided over by a skeleton with averted gaze to which, paradoxically, the eye is inexorably drawn. The corpse is indecently jostled by a crowd who from their dress and various attitudes (one consulting a text, one in an attitude of prayer, others sitting on the floor preoccupied with dogs, monkeys and other distractions, with two naked, bearded, godlike figures peering from behind the pillars) are as various both in class and in intent as any theatre audience of the day. There is an extraordinary sense of intrusion and prurience implicit in the scene, pointed up by the central skeleton figure; it is as if Death cannot bear to see its mysteries broached in the manner of an open public entertainment. Sawday further points out how 'opposites are brought face to face in the anatomy theatre; the spontaneous with the staged, the living with the dead, knowledge with ignorance, civic virtue with criminality, and judicial power with individuality.' And let's be clear too how swiftly Vesalius' influence was felt across Europe. Only six years after the publication of *De humani corporis fabrica* the statutes of Oxford University laid down that medical students must see at least two dissections during the course of their studies, and by 1565 Gonville and Caius College, Cambridge was taking two bodies a year, the Royal College of Surgeons, four.

9 For further detail on this see Rice and Grafton, pp 22–6. **10** Rice and Grafton, p. 24. **11** See Jonathan Sawday, *The body emblazoned: dissection and the human body in renaissance culture* (London and New York, 1995).

So why exactly was Vesalius' text – like Copernicus' of the same year – so revolutionary? As a student, Vesalius and his aptly named assistant Matthew Terminus had been mightily frustrated by the difficulty of gathering, from among the children's graves at the Cemetery of the Innocents, enough good specimens to construct a fully articulated skeleton. Frustrated himself in attempts to work properly from life, he was determined to bring to press a clearly illustrated and articulated account of what was the standard text of anatomy, Galen's *On the usefulness of the parts of the human body* in a new, universally accessible Latin version. Galen had written in the second and third centuries AD, in Greek, then the most respected language in medicine, but Latin was the universal language of Renaissance Europe; Greek was only just beginning to be taught again in English universities in the sixteenth century. Galen's texts, based in part on his experience as physician to gladiators in Pergamum had recently become available in their original form in Western Europe because when the Turks took the city of Byzantium from western Christendom in 1453, Greek scholars fled to the West, many of them bringing precious texts with them. Vesalius, in reworking Galen, made a number of substantial changes. Where Galen had often worked from animal vivisection, Vesalius worked from the dead human body (a Paduan judge became interested in his work in 1539, and agreed to make the bodies of hanged felons available for dissection). He took a huge amount of care in the production of the text, having the prestigious studio of Titian produce the images, and the best printing house in Europe, at Basel, responsible for the printing. Vesalius' instructions to the printer, presented as part of the text itself, describe the relationship of the text to its originals, protests against earlier accusations of lack of originality, goes into extraordinary detail abut how important the clarity of the reproduction of the images are, and shows exactly how important both the craftsmanship of the engraver and the new technology of print were to the accurate production of new knowledge:

> Special attention will have to be paid while printing the plates, because they are not just simple outlines drawn in the common schoolbook manner; the artistic style (except sometimes where the surface is outlined on which the specimens are resting) is nowhere neglected [...] I have the highest expectations regarding your meticulous craftsmanship [...] no identifying character, however hidden in shading, will escape the sharp-eyed and careful reader [...] but I should not write these things out for you, since it depends on the smoothness and solidity of the paper, and particularly on the diligence of your efforts[12]

12 Vesalius, *De humani corporis fabrica*, pp vii–viii. http://vesalius.northwestern.edu/noflash.html, accessed 2 August 2006.

Technology aside, there was a fundamental methodology emerging here too. In the dedication to Charles V, Vesalius announces 'I decided that this branch of natural philosophy ought to be recalled from the region of the dead',[13] and it is clear that this is more than just a rhetorical flourish. He resurrects the work of Galen, of course, but he also places his subjects in extraordinary positions, skeletons leaning on spades in a barren landscape, skeletal jaw agape in what appears to be a howl of pain or standing at a tomb in contemplation of a skull; a flayed body with its skin hanging in shreds and all its musculature displayed (and neatly marked 'a, b, c' as per the instructions to the printer) gazing longingly back towards a town half-hidden behind hills, symbolic, one supposes, of the life just left behind by the corpse.[14] The bodies are given a context; the study of the skeleton and of the musculature does not exist in isolation from remembrance of the fact that these objects have also a social existence. The notion that the figures have artistic as well as medical function is an extension of the same paradigm: that practicality and beauty, physical functionality and spirituality, commodification and social status, are complementary aspects of what it is to be human.

What Vesalius does visually he also does verbally. By describing anatomy as a branch of natural philosophy, he gives it a higher status than it has enjoyed until now. Because anatomy was to do with the mechanical workings of the human body, rather than the holistic wellbeing of a person – because it dealt with the body as separate from the spirit – it had a much lower status than other branches of medicine. But the name by which Vesalius calls his study, natural philosophy, in the scholastic tradition deals with causes of things, reasons for them, rather than simple descriptions of them. We don't just need to know what a body is, or how it looks, or how it functions; we need to know *why* it is the way it is, Vesalius suggests, perhaps even how it came to be the way it is.

Like Copernicus, Vesalius had to embed his new thinking in a system supported by ancient authority. It was slightly more difficult for Copernicus; he was not urging a change in status for his subject, but an absolute change in thinking. In his introduction to *De revolutionibus*, Copernicus says:

> I undertook the task of rereading all the philosophers which I could obtain to learn whether anyone had ever proposed other motions of the universe's spheres than those propounded by the teachers of astronomy in the schools. And in fact I first found in Cicero that Hicetas supposed the earth to move. Later I also discovered in Plutarch that certain others were of this opinion [...] therefore I too began to consider the mobility of the sun around the earth.[15]

13 Translated in C.D. O'Malley, *Andreas Vesalius of Brussels 1514–1564* (Berkeley, 1964) p. 320. Cited in Dear, p. 38. **14** These images are available at the website cited, and a selection of them is contained in Michael Neill's *Issues of death: mortality and identity in English renaissance tragedy*. **15** Nicholas Copernicus, *On the revolutions*, trans. and commentary by Edward Rosen (Baltimore, 1992) p. 4. Cited in Dear, p. 35.

He later uses a quotation from Virgil's *Aeneid* to explain how one might feel stationary on a moving earth, just as one might feel stationary on a moving ship. It is interesting that ocular proof is not enough, as Galileo was to learn to his cost; but by 1596 Galileo can say in a letter to Kepler that he is a Copernican – that is, his premises are based on a mathematical calculation coupled with observation. And by 1618 Kepler can write that astronomy is 'a a part of physics [that is, natural philosophy] because it seeks the causes of things' and that the 'astronomer directs all his opinions, both by geometrical and physical arguments, so that truly he places before the eyes an authentic form'. The shift here is in the cataloguing of the kind of research being done. Philosophy – the working out of things through rhetorical logic – is beginning to lose ground to physical observation, and these works can be disseminated and built upon because exactly consonant with the work comes the development of the printing press. This, especially in Germany, allows for the printing of books of practical craft, texts which had been of no interest to monastic or university scribes since the practicalities of individual things had held little interest in the university curriculum, which had consisted, for centuries, of grammar, logic and rhetoric, followed by arithmetic, mathematics, music and geometry, subjects that were believed to provide the perfect foundation for proceeding to a vocational course in law, divinity, or medicine.

The point here is that new science had to establish itself, justify itself against the old systems, and find legitimate ways of disseminating itself. It had to take cognisance of what had gone before it, and because the humanities had held sway, to negotiate its position among the traditional methods of practice and justify its usefulness to the understanding of humankind. So if this is how the sciences are informed by the arts, how are the arts informed by the sciences? This question forms the basis of the next section of this essay.

* * *

The Trinity College Dublin Early Printed Books catalogue yields up some fifty-eight works with the word 'anatomy' or one of its cognate forms in the title. These include the anatomy lectures at Gresham College, the anatomy of a porpoise, earthworm, horse, opossum, of plants, of the brain, of human bodies, of the body of man, of the bodies of men and women; the rather unfortunate 'anatomy of a pygmy compared with that of a monkey, an ape and a man'; less obviously, for the scientific mind, perhaps, there are works referring to the anatomizing of tyrants, the anatomy of a Christian man, of the abuses in England, of Simon Magus, of a Jacobite Tory, of an arbitrary prince, of Armenianism, of atheism, of consumption, of Dr Gauden's idolized nonsense and blasphemy, of Calvinistical calumnies, of England's vanities, of fair writing, of labour, of independency, of legerdemain, of melancholy, of money, of play, of popery, of secret sins, of the English nunnery at Lisbon in Portugal, of the 'late unhappy mutiny'. Anatomies of the mass, of the service book, of the world, of transubstantiation, of war are there too, and my two personal favorites – 'An anatomy of et cætera' and, most fantastically, 'The anatomy of the separatists, alias, Brownists; the factious brethren in these times, wherein this seditious sect is fairely dissected, and perspicuously discovered to the view of the world; with the strange hub-bub and formerly unheard of hurly-burly, which those phanatick and fantastick schismaticks made on Sunday the 8th of May at the sermon of the right reverend father in God, Henry, Bishop of Chichester.' It is evident in this list that the notion of taking something apart to examine it minutely has transferred across a whole series of disciplines, sometimes in a literal sense ('the anatomy of human bodies epitomised') and sometimes in a metaphorical one, as we saw in the last. If we turn to drama, or poetry, rather than political tracts, what we find is a similar shift from the literal to the metaphorical and back again.

Mankind, a morality play written about 1465, makes a clear distinction between the physical and spiritual sides of humankind's condition: this is how the character of 'Mankind' introduces himself on stage:

> Of the earth and of the clay we have our propagation;
> By the providence of God thus be we derivate –
> To whose mercy I recommend this whole congregation;
> I hope unto his bliss ye be all predestinate.
> Every man for his degree I trust shall be participate,
> If we will mortify our carnal condition
> And our voluntary desires, that ever be perversionate,
> To renounce them and yield us under God's provision.
> My name is Mankind. I have my composition
> Of a body and of a soul, of condition contrary.
> Betwix them twain is a great division;
> He that should be subject, now he hath the victory.
> This is to me a lamentable story,
> To see my flesh of my soul to have governance. (186–99)[16]

Great play – both highly comic and deeply tragic – is made of the temptations of the flesh in the play, and it is much more complex in exposition and in its ability to move emotionally than I have time to explore here. It is interesting though that one of the earliest plays we have in English that is not directly aligned with a Bible

16 Anon., *Mankind*, in G.A. Lester (ed.), *Three late medieval morality plays* (1981; rpt. London, 1990). The standard edition is in Mark Eccles. ed., *The Macro plays* EETS no. 262 (London, 1969) where the play is given in original spelling. I have used a modern spelling edition here in the interests of accessibility.

story chooses to introduce the figure of humankind in this way, as caught between the flesh and the spirit.

A slightly later moral play, *A Play of Love*, is vastly more irreverent and deliberately complicated, and it investigates the notion of anatomy in a number of ways.[17] The play sets up a debate among four characters – Lover-not-loved, Loved-not-loving, Lover-loved and No-lover-nor-loved – to find out by logical debate and experiment which of the four conditions is least and most painful. A whole series of systems of belief are exploded in the course of the play (the Petrarchan opposition of heat and cold, and of fire and water, quite literally at two separate moments) and of course we are brought to realize at the end that the only perfect love is the love of Christ for humankind; but two or three sections of the play are especially interesting for our purposes here. One is where the idea of Petrarchan opposition in poetry – Petrarch regularly describes the pains of love in terms of icy fire, loving hatred, darkness during the daytime – is introduced, but then immediately aligned with a physical, pseudo-medical examination of the lover, who is claiming to be in pain because of unrequited love. Dismissing this pain as a poetical fancy, No-lover-nor-loved conducts his investigation according to Galen's theory of the four humours, whereby the body is governed by a series of groups of four; anything that is wrong with you can be cured by rebalancing the imbalance between these humours in your body. So, at a very basic level, if you are angry, that means you are choleric, which means you are too hot and dry, so you should go off and do something wet and warm, and you will soon feel better. Here is what happens in *A Play of Love*: Lover-not-loved is moaning in a fairly hackneyed way, like the typical Petrarchan lover, and No-lover-nor-loved tries to sort him out by showing that Petrarchan opposition is only a notional, not a physical way of being:

> Loving-not-loved*: All time in all places distempered am I;*
> Shivering in cold, and yet in heat I die;
> Drowned in moisture, parched parchment dry.
> *No-lover-nor-loved*: Cold, hot, moist, dry in all places at once?
> Marry sir this is an ague for the nonce.
> But for we give judgement I must search to view
> Whether this evidence be false or true.
> Nay, stand still your part shall prove never the worse;
> So, by saint Saviour, here is a hot arse!
> Let me feel your nose – nay fear not man, be bold,
> Well, tho' this arse be warm and this nose be cold
> Yet these twain by attorney brought in one place
> Are as he saith cold and hot both in like case. (1021–35)

17 John Heywood, *A Play of Love*, in J.A.B. Somerset (ed.), *Four Tudor Interludes* (London, 1974).

Out-of-date poetical rambling must be subjected to a serious medical test or 'search' – the verb generally used when a surgeon removes the injurious object from a wound, is the giveaway term here. Because the practical, objective test according to the four humours is demonstrated to be right, rather than simply described as right, as the subjective, poetical self-diagnosis is, it is the latter which is held to be more acceptable. Except, of course that it is all done in bawdy fun, so that what we are actually left with is a mocking of the notion that medicine gives any more real a diagnosis in affairs of the heart than poetry might. Systems of belief, whether they are poetic or practical, whether they use the same language or not, are treated sceptically in the play.

More straightforward in the play is the top-to-toe anatomy of the physical prompts to love as No-lover-nor-loved tells how he fell in love but was cured of it, describing his former lover in the following terms:

For to begin
At setting in
First was her skin
Hite smooth and thin
And every vein
So blue seen plain
Her golden hair
To see her wear
Her wearing gear –
Alas I fear
To tell all to you
I will undo you.
Her eye so rolling
Each heart controlling
Her nose not long
Nor stood not wrong
Her finger tips
So clean she clips
Her rosy lips
Her cheeks gossips
So fair so ruddy
It asketh study
The whole to tell
It did excel
It was so made
That even the shade
At every glade

Would hearts invade
The paps as small
And round withal
The waist not mickle
But it was tickle
The thigh, the knee
As they should be
But such a leg
A lover would beg
To set eye on –
But it is gone.
The sight of the fote
Rift hearts to the rote
And last of all
St Catherine's wheel
Was never so round
As was her heel. (429–70)

This is a dutiful tour around the body; in the context of the way the play works it is part of the parody of assuming that real knowledge of anything can come either through physical knowledge or rhetorical knowledge. Like the anatomist, the playwright knows that however thorough your description, there will always be a particular element that remains unidentifiable.

Shakespeare's *The Comedy of Errors* (1594) has an extended anatomical moment too, but in this case it is an extended parody of the body as the little world, as Dromio explains to his master why he finds amorous advances of a local kitchen maid unattractive, and Antipholus amuses himself by seizing on his servant's first observation ('She is spherical, like a globe') and using it as the pattern for the rest of the exchange:

Dromio of Syracuse: She is spherical, like a globe. I could find out countries in her.
Antipholus of Syracuse: In what part of her body stands Ireland?
Dromio: Marry, sir, in her buttocks. I found it by the bogs.
Antipholus: Where Scotland?
Dromio: I found it by the barrenness, hard in the palm of her hand.
Antipholus: Where France?
Dromio: In her forehead, armed and reverted, making war against her heir.
Antipholus: Where England?

Dromio:	I looked for the chalky cliffs, but I could find no whiteness in them. But I guess it stood in her chin, by the salt rheum that ran between France and it.
Antipholus:	Where Spain?
Dromio:	Faith, I saw it not, but I felt it hot in her breath.
Antipholus:	Where America, the Indies?
Dromio:	O sir, upon her nose, all o'er embellished with rubies, carbuncles, sapphires, declining their rich aspect to the hot breath of Spain who sent whole Armadas of carracks to be ballast at her nose.
Antipholus:	Where stood Belgia, the Netherlands?
Dromio:	O sir, I did not look so low. To conclude, this drudge or diviner laid claim to me, called me Dromio, swore I was assured to her, told me what privy marks I had about me – as the mark of my shoulder, the mole in my neck, the great wart in my left arm – that I, amazed, ran from her as a witch.(III. ii. 112–42)[18]

Dromio's comment is initially intended to describe shape and bulk, but between them the two men show how easy it is to construct a rhetorical system out of a single observation, and how that system becomes pretty much self-perpetuating, only brought up short here by Dromio's reluctance to descend beyond a certain point in the description. What is clearly being parodied here is the idea that to understand the body by dividing it into its parts is to understand the world – that the anatomizing of humankind is absolutely transferable as a metaphor for the understanding of the whole world. Here it is in parody, but in poems of John Donne, with which I want to conclude, we see it taken very seriously.

* * *

The notion of anatomy as a new world of discovery is celebrated in Donne's famous 'Elegy XIX. To His Mistress Going to Bed', where he writes: 'Licence my roving hands and let them go. / Before, behind, between, above below' and describes his mistress as 'my America, my new-found-land, / My kingdom, safeliest when with one man manned'.[19] Incidentally, Donne was refused a licence to print this poem in 1633, but his poem entitled 'An Anatomy of the World' (1611) takes apart the notion of anatomy as a way of knowing pretty much everything there is to be known – except what happens after death. The poem is written to

18 William Shakespeare, *The Comedy of Errors*, in Stanley Wells and Gary Taylor (eds), *The Oxford Shakespeare: the complete works* (2nd ed. Oxford, 2005), pp 287–305. **19** John Donne, *Poetical Works*, ed. Herbert J.C. Grierson (Oxford, 1933), pp 106–8.

mark the death of a young girl, perhaps a family friend of Donne's, perhaps not. It is a long poem and we can only look at a small part of it here:

> But though it be too late to succour thee,
> Sicke world, yea, dead, yea putrified, since shee
> Thy'intrinsique balme, and thy preservative
> Can never be renewed, and never live,
> I (since no man can make thee live) will try,
> What we may gain by thy Anatomy.
> Her death hath taught us dearely, that thou art
> Corrupt and mortall in thy purest part.
> Let no man say, the world itselfe being dead,
> 'Tis labour lost to have discovered
> The worlds infirmities, since there is none
> Alive to study this dissection;
> For there's a kinde of World remaining still,
> Though shee which did inanimate and fill
> The world, be gone, yet in this last long night
> Her ghost doth walke; that is, a glimmering light,
> A faint weake love of virtue, and of good
> Reflects from her, on them which understood
> Her worth; and though she have shut in all day,
> The twilight of her memory doth stay;
> Which, from the carcasse of the old world, free,
> Creates a new world; and new creatures be
> Produc'd; the matter and the stuff of this
> Her vertue and the forme our practice is […] (55–78)[20]

The metaphor here concerns all knowledge; that although old knowledge is dead new knowledge springs from it, just as the anatomy lesson can teach us, from a corpse, things about the living body. Even if the best part of humankind vanishes at death, there is still something left from which we can learn.

The poem goes on to describe how humankind's command of knowledge has come to seem small, since knowledge of the world has expanded – since we know that there is so much more to be known, and how, as he writes:

> […] new Philosophy calls all in doubt
> The element of fire is quite put out;
> The Sun is lost, and th'earth, and no man's wit
> Can well direct him where to looke for it,

20 Ibid., pp 206–21; p. 209.

And freely men confesse that this world's spent,
When in the Planets and the Firmament
They seeke so many new; they see that this
Is crumbled out againe to his Atomies.
'Tis all in peeces, all cohaerence gone;
[..............................]
We thinke the heavens enjoy their Sphericall,
Their round proportion embracing all
But yet their various and perplexed course,
Observ'd in divers ages, doth enforce
Men to finde out so many Eccentrique parts,
Such divers down-right lines, such overthwarts
As disproportion that pure form …
[..............................]
Shee, shee is dead, shee's dead; when thou know'st this
Thou knowst how ugly a monster this world is;
And learn'st thus much by our Anatomie,
That here is nothing to enamour thee … (203–16, 251–7, 325–8)[21]

In the final passage quoted here Donne laments the passing of innocence in the death of the young girl, and uses the idea of the loss of innocence as an extended metaphor for a world in chaos. The tight control of the form of the poem however belies the chaos described; there is a tension here between what is being said and the way in which it is being said. When the poem is searched carefully, it reveals layer upon layer of meaning, just as the anatomist reveals function after function, wonder upon wonder, as he proceeds fearfully and, as Vesalius' illustrations and especially the frontispiece to *De corporis* show, with a sense of intrusion, upon his task of dissection. The project of anatomy can reveal a detailed understanding of the smallest part of the human body; but it can also overturn humankind's most fundamental understandings of itself because of the questions it does not answer. It can stand as a series of factual, objective observations on the nature of the machine that is the human body, but it can also stand as a representation of that final, fundamental, unfathomable question: what is it that makes us what we are? Where anatomy is concerned the projects of science and of art speak mutually to each other, but each realizes that the process of discovery involves losing as much as is gained. Barlaeus' notion that, when 'that which is hidden in our bodies' fabrica is brought to light […] whatever we were, we are no longer' summarizes perfectly the loss of poetic mystery and spiritual ineffability that necessarily attends the laying bare of what has been secret.

21 Ibid., pp 213–14; pp 215, 217.

Broken machines and tainted minds: mental health and *Hamlet*

ANDREW J. POWER

I: IS THERE A DOCTOR IN THE PLAYHOUSE?

Over the years many critics have attempted to find an appropriate diagnosis for Hamlet's apparent mental distemper. Irving I. Edgar has traced the four main interpretations of Hamlet's character that have been forwarded:

> 1. that the character represents a true feigned psychosis; 2. that it represents a combination of feigned psychosis and actual psychosis; 3. the historical interpretation which holds that Hamlet portrays the melancholy type common in Elizabethan literature and Elizabethan psychology; and 4. that the character represents a true example of an unresolved Oedipus complex.[1]

In 1999 Robert M. Youngson added another. He suggested that Hamlet suffers from Ganser syndrome. This syndrome, he explained, is 'the voluntary production of psychiatric symptoms'.[2] This, however, seems simply to give a modern technical name to putting on 'an antic disposition' (I. i. 180).[3] He admits that psychiatrists find it hard 'to make up their minds as to whether Ganser syndrome is a genuine psychiatric entity or just another all-too-human ploy in the battle for personal survival in a hostile world.'[4]

The attention, though largely unfruitful, that has been given to the problem of Hamlet's sanity is not surprising when we consider that much of the first half of the play is devoted to the attempt, by the leaders of the Danish court, to discover the nature of Hamlet's mental anguish. Polonius is the first to venture an explanation for his condition:

1 Irving I. Edgar, *Shakespeare, medicine and psychiatry: an historical study in criticism and interpretation* (London, 1971), p. 255. 2 Robert M. Youngson, *The madness of Prince Hamlet and other extraordinary states of mind* (London, 1999), p. 1. 3 All quotations from the play, unless indicated otherwise, are from William Shakespeare, *Hamlet*, ed. Harold Jenkins, the Arden Shakespeare (London, 1995). 4 Youngson, p. 3.

> This is the very ecstasy of love,
> Whose violent property fordoes itself
> And leads the will to desperate undertakings
> As oft as any passion under heaven
> That does afflict our natures. (II. i. 102–6)

The king has invited Rosencrantz and Guildenstern to Denmark to find out more for him, though he himself suspects it is simply grief,

> Something have you heard
> Of Hamlet's transformation – so call I it,
> Sith nor th'exterior nor the inward man
> Resembles that it was. What it should be,
> More than his father's death, that thus hath put him
> So much from th'understanding of himself
> I cannot dream of [...] (II. ii. 4–10)

And yet he is understandably suspicious that it might be something else and bids them 'gather, / So much as from occasion you may glean, / Whether aught to us unknown afflicts him thus / That open'd, lies within our remedy.' (II. i. 15–8) Gertrude quite reasonably asserts that her 'too much changed son' (II. ii. 36) suffers from 'no other but the main, / His father's death and [her] o'er-hasty marriage.' (II. ii. 56–7) Neither are the observations of Rosencrantz and Guildenstern the only means used to investigate the cause of his condition. Convinced that it is the ecstasy of love, Polonius determines to 'board him' (II. ii. 170) in an attempt to discover more. But all he can glean is that Hamlet's replies are sometimes cryptic and yet somehow meaningful, 'a happiness that often madness hits on, which reason and sanity could not so prosperously be delivered of.' (II. ii. 208–11) Hamlet himself admits that he cannot explain what is wrong, 'I have of late, but wherefore I know not, lost all my mirth, forgone all custom of exercises; and indeed it goes so heavily with my disposition that this goodly frame the earth seems to me a sterile promontory'. (II. ii. 295–9)

When Gertrude, Claudius, Polonius, Rosencrantz, and Guildenstern have all failed in the attempt to discover the nature of his apparent illness even Ophelia is consulted in the case. From her examination of the patient she has nothing useful to add:

> O, what a noble mind is here o'erthrown!
> [..............................]
> Now [I] see that noble and most sovereign reason
> Like sweet bells jangled out of tune and harsh,

That unmatch'd form and feature of blown youth
Blasted with ecstasy. (III. i. 152; 159–62)

It is thus not surprising that critics have enlisted the help of psychoanalysis, psychiatry and contemporary melancholy theory in an attempt to uncover the nature of Hamlet's malady. What is surprising is that in the play the opinion of a doctor is never sought. Doctors are consulted for the mental extremities of King Lear,[5] Lady Macbeth,[6] and the Jailer's Daughter.[7] Dr Pinch is consulted for the apparent lunacy of Dromio and Antipholous, and a mock doctor, Sir Topaz, is brought to examine Malvolio.[8] If not for Hamlet himself, the Danish court's failure to avail of the services of a doctor for Ophelia's more obvious mental anguish is hard to excuse. Kenneth Branagh's movie version went so far as to give her a padded cell despite the lack of any textual evidence to suggest any medical attention whatsoever.[9]

If the absence of a doctor in the play is unusual, it would not necessarily be unusual in early-modern society to consult other professionals as an alternative to, or in conjunction with, a doctor. In *Twelfth Night* the professional called upon to visit Malvolio in his madness is a cleric, or at least so he seems. Feste, the fool, disguises himself as 'Sir Topaz the curate' (*Twelfth Night*, IV. ii. 2) to visit Malvolio in his dark room. Though the incident is humorous, though Malvolio is not in fact mad and is thus wrongly confined (for his treatment) in a dark room, and although Feste is no cleric, it does point to an acceptable contemporary alternative to consultation with a doctor in cases of madness. Some of the most commonly referenced medical publications of the period were written by priests or by authors closely connected to the church. Timothy Bright, who became rector of Methley in 1591, was the author of *A treatise of melancholie* (1586) often cited as a source for Hamlet's characteristic melancholy;[10] Thomas Wright, a Catholic priest imprisoned for his religious publications also wrote *The passions of the mind in general* (1601);[11] M. Andreas Laurentius, the author of *A discourse of the preservation of sight: of melancholike diseases; of rheumes, and of old age* (1599) had two brothers who were made archbishops in the Catholic church.

There is a priest in *Hamlet*. His appearance, however, comes too late to be of any assistance to Ophelia or Hamlet and he is described by Laertes as a 'churlish priest'. (V. i. 232) His one significant pronouncement in the play is that Ophelia

5 William Shakespeare, *King Lear*, ed. R.A. Foakes, the Arden Shakespeare (Surrey, 1997) IV. viii. 6 William Shakespeare, *Macbeth*, ed. Kenneth Muir, the Arden Shakespeare (Surrey, 1984) V. i. 7 William Shakespeare and John Fletcher, *The Two Noble Kinsmen*, ed. Lois Potter, the Arden Shakespeare (Surrey, 1997) IV. iii. 8 William Shakespeare, *The Comedy of Errors*, ed. R.A. Foakes, the Arden Shakespeare (London, 1962) IV. iv., and William Shakespeare, *Twelfth Night*, eds J.M. Lothian and T.W. Craik, The Arden Shakespeare (London, 2000) III. iv. 9 Kenneth Branagh, dir. *William Shakespeare's Hamlet*, Castle Rock Entertainment, 1996. 10 Mary I. O'Sullivan's 'Hamlet and Dr. Timothy Bright' (*PMLA* 41 [1921], 667–79) offers a comprehensive comparison of the two texts. 11 A manuscript copy of Wright's text was circulated as early as 1597.

must not be given full burial rites. This is in spite of the fact that, as the second clown, companion to the gravedigger, says, 'The crowner hath sate on her and finds it Christian burial.' (V. i. 4–5) Ophelia is a suicide: her entitlement even to these 'maimed rites' (V. i. 212) is due to the intervention of the 'great command' of court authority. (V. i. 221) The services of a medical professional, though a corruptible coroner, are employed for the dead but not for the living as are the services of a priest, but without their promise of succour, medical or otherwise. A mountebank also serves one of the characters in play, but it is not to cure. The poisonous unction with which Laertes anoints his blade for the duel was bought off a mountebank. (IV. vii. 140)

Andrew Boorde suggests a further alternative for the treatment of the mad and the sick. He is himself the author of a dietary, and says, 'A good cook is half physician. For the chief physic [...] doth come from the kitchen; whereof the physician and the cook must consult together for the preparation of meat for sick men.'[12] There is, however, no sign of a cook or dietician in *Hamlet*, just the 'funeral bak'd meats' that 'coldly furnish forth the marriage tables.' (I. ii. 180–1) Finally, and in more extreme cases, a barber surgeon might be asked to perform more radical surgery, such as bloodletting or amputation.

When Guildenstern comes to reprimand Hamlet for his mischievous play, he tells him that the king is 'marvellous distempered [...] with choler.' (III. ii. 293, 295) Hamlet, though in jest, points to the curious absence of professional medical attendance within the court: 'Your wisdom should show itself more richer to signify this to the doctor, for for me to put him to his purgation would perhaps plunge him into more choler.' (III. ii. 296–9) There is blood let in the play, but there is unfortunately no sign of a barber other than Hamlet's threat to Polonius, 'It shall to the barber's with your beard.' (II. ii. 495) In the absence of any of the above-mentioned 'qualified' members of society who might under ordinary circumstances have been of assistance to the diseased minds of Denmark, the characters themselves are left to pronounce diagnosis and to attempt to administer treatment of their own accord.

A humourous disposition

All illness in the early modern period was explained by reference to humoural physiognomy, which was as basic to early-modern physiological knowledge as the circulation of the blood is to our own. Galen, whom Andrew Boorde describes as the 'prince of phisicions', attributes the origins of the theory to Hippocrates, who described the body of man as containing four humours. These are blood, choler, yellow bile, and black bile.[13] The health of the body depends upon the balance of

12 Excerpts from Andrew Boorde's *Dietary of health* are included in H. Edmund Poole (ed.), *The Wisdom of Andrew Boorde* (London, 1936), p. 49. **13** The word 'blood' is used both as a general term for the fluid that contains all four humours and as the humour prevalent in the sanguine constitution.

these four humours. Excessive amounts of any one humour will cause a sickness with the characteristics of that humour.[14] Hamlet mentions such an excess – an 'o'ergrowth of some complexion' that may break down 'the pales and forts of reason'[15] – as being comparable to men who have a tendency towards a single sin, 'a vicious mole of nature in them [...] one defect' that causes their complete corruption. (I. iv. 23–38) His association of sickness and moral guilt is characteristic of the medical authorities but it is also significant that he believes the abundance of a humour to be capable of undermining reason.

What Hamlet specifically refers to is the reputation that the Danes have for drinking, an overindulgence which, while it is a sin of gluttony, might also cause humoural imbalance by its effect on digestion. As Bridget Gellert Lyons explains, 'According to Galenic thought, blood (and the humours which composed it) was manufactured from food, whose properties determined which of the humours would be produced in greatest quantity.'[16] The stomach is the first organ involved in digestion performing the first 'concoction'. It takes the food and breaks it down into the substance known as chyle. This substance, chyle, is then passed to the liver, which Boorde refers to as the 'fire under the pot' where the second concoction takes place.[17] This second phase of digestion breaks the chyle down into the four humours. Food and drink are thus of the utmost importance in the maintenance of a healthy body. Most illnesses were considered to be caused by bad diet, and likewise most were possible to cure by an emendation to the diet that would duly balance the humours. Nobody recommends a dietary change for Hamlet.

The nature of the soul

The reason that the sanguine temperament and the humour blood are so highly prized is that they contain the two main qualities of life, heat and moisture. For this reason blood is corresponded to youth and childhood. But these qualities do not last forever and as we age these qualities decline in us so that by the time we are old we become cold and dry like melancholy. This heat and moisture comes from the soul which we are endowed with at birth by the breath of God. 'It is created of God, and commeth downe from heauen to gouerne the bodie, so soone as the members thereof are made: the effects thereof doe sufficiently proue vnto vs, the worthines of the same.'[18] The soul is for practical purposes one soul, but it was also thought by some commentators to be comprised of three separate souls. The disagreement is twofold. The first problematic aspect arises over whether the soul of man is mortal or whether it survives after the body dies.[19] The early med-

14 Boorde, pp 12–13. **15** Hamlet's reference to the 'forts of reason' is reminiscent of Bright's description of reason's 'watch tower'. See Timothy Bright, *Treatise of melancholie* (1586), intro. Hardin Craig (New York, 1940), 'Epistle dedicatory', n.p. See also O'Sullivan, p. 669. **16** Bridget Gellert Lyons, *Voices of melancholy: studies in literary treatments of melancholy in renaissance England* (London, 1971), p. 2. **17** Boorde, p. 36. **18** M. Andreas Laurentius, *A discourse of the preservation of sight: of melancholike diseases; rheumes, and of old age*, trans. Richard Surphlet (1599), intro. Sanford V. Larkey (London, 1938), p. 73. **19** E. Ruth Harvey gives

ical treatise writers, Galen and Avicenna in particular, follow the Stoics who believed in a *pneuma*, which was a physical substance and which roughly equated to a soul.[20] In contrast, Platonic, and Christian thinking, distinguished between body and soul, the soul being 'a divine and deathless incorporeal substance'.[21]

It is perhaps inevitable that when Christian writers engage with the matter of the soul in medical treatises there should be some debate. As Christians, they were bound by their faith to believe in the soul's immortality, but as physicians they had to reconcile this with the medical theories that had until recently accounted the soul a substantial thing. Bright proposes to address this mistake of the earlier physicians, saying of medicine's success in curing people,

> The notable fruit & successe of which art in that kinde, hath caused some to iudge more basely of the soule, then agreeth with pietie or nature, & haue accompted all maner affections thereof, to be subiect to the physicians hand, not considering herein any thing diuine, and aboue the ordinarie euents, and naturall course of thinges: but haue esteemed the vertues them selues, yea religion, no other thing but as the body hath ben tempered, and on the other side, vice, prophanenesse, & neglect of religion and honestie, to haue bene nought else but a fault of humour [...] I haue layd open howe the bodie, and the corporall things affect the soule, & how the body is affected of it againe ...[22]

The second element over which the medical writers disagree involves the number of souls a body contains. The souls of humans were considered to serve three discrete spheres of operation: vegetative, sensitive, and rational. The debate was nominal rather than substantial. These three elements were at times described as taking place in the vegetative soul, the sensitive soul, and the rational soul. Robert Burton attempts to reconcile objections to this apparently sacrilegious suggestion that there were three souls rather than one with reference to previous medical authorities including Hippocrates, Galen and Avicenna. He concludes, making explicit his belief that the soul is eternal, 'all three Faculties make one *Soule*, which is inorganicall of it selfe, although it be in all parts, and incorporeall, using their Organs, and working by them.'[23] Effectively, the three spheres of operation are due to bodily distinctions rather than spiritual ones.

Bright attempts to reconcile these problems and to explain the movement of the earthly, corruptible, and intrinsically inert flesh of the body when he likens it

a detailed account of the varying arguments over the immortality of the soul in *The inward wits: psychological theory in the middle ages and the renaissance* (London, 1975), pp 31–48. **20** Harvey, p. 31. **21** Harvey, p. 32. **22** Bright, 'Dedicatorie Epistle', n.p. **23** Robert Burton, *The anatomy of melancholy* (1621), 3 vols, ed. Thomas C. Faulkner, Nicholas K. Kiessling, and Rhonda Blair, intro. J.B. Bamborough (Oxford, 1989–94) i, pp 155–7.

to a complex clock. The clock may seem to work on its own, to point with hands to the time, to 'counterfet voice' and declare the time with noise (by striking a bell), and in some instances 'you may finde all the motion of heauen with his planets counterfetted, in a small modill, with distinction of the time & season, as in the course of the heauenly bodies.'[24] This animation is not inherent in the machine, however, but is given to the machine by whomever first winds the spring. 'So many actions diuerse in kinde rise from one simple first motion, by reason of variety of ioynts in one engine.'[25] The machine corresponds to the human body, and the 'one simple first motion' is the soul, the 'breath of life [...] inspired from God'.[26] This passage bears a notable resemblance to the love letter from Hamlet that Ophelia shows to her father, as described in this exchange:

Polonius: I have a daughter – have while she is mine –
Who in her duty and obedience, mark,
Hath given me this. Now gather and surmise.
(*Reads*) *To the celestial and my soul's idol, the most*
beautified Ophelia – that's an ill phrase, a vile phrase,
'beautified' is a vile phrase. But you shall hear –
these; in her excellent white bosom, these, &c.
Queen: Came this from Hamlet to her?
Polonius: Good madam, stay a while, I will be faithful.
Doubt that the stars are fire,
Doubt that the sun doth move,
Doubt truth to be a liar,
But never doubt I love.
O dear Ophelia, I am ill at these numbers. I have not the art to
reckon my groans. But that I love thee best, O most best,
believe it. Adieu.
Thine evermore, most dear lady, whilst this
machine is to him, *Hamlet.* (II. ii. 107–125)

The second line of the brief and cryptic quatrain read by Polonius casts doubt on the progress of the sun, 'Doubt that the sun doth move'. That the Ptolemaic (earth-centred) system has come into doubt by the time Shakespeare wrote the play adds irony to the verse, but within the context of the play, if the sun stops moving, then time cannot be measured (as it might be on a sundial).[27] When

24 Bright, pp 67–8. **25** Ibid., p. 68. **26** Ibid., p. 37. **27** In 1543 Copernicus published *De revolutionibus orbium coelestium* (*On the revolution of the celestial spheres*) propounding the theory that we live in a heliocentric universe (i.e. that the earth and all the planets revolve around the sun). See Jon Balchin, *Quantum leaps: 100 scientists who changed the world* (London, 2003), pp 38–9. Timothy Bright (writing in 1586) was convinced of the correctness of Copernican theory. Within the context of Shakespeare's play, Hamlet is of

Hamlet says he is 'ill at these numbers', the immediate sense is that he cannot compose his verse within the meter. But it also means simply that he cannot 'count', a sense echoed again in the word 'reckon'. Again, if he cannot count, he cannot keep time. His ending seems initially strange. What is 'this machine'? It suggests that his body is the machine, the elaborate mechanism in which his soul is for a limited space of time housed. So, he means, 'as long as I am alive, as long as my soul is housed in this body, I am yours.'[28] Bright figures the human body as a clock, 'an engine' he calls it. Hamlet calls his body a 'machine'. Bright says that by striking a bell periodically a complex clock may tell the time; Hamlet cannot reckon (or count) his groans. Bright says that if the clock is complex enough it may present all the motions of the planets and heavenly bodies; Hamlet doubts 'that the sun doth move'. Hamlet is cryptically alluding to a bodily malfunction; his machine, his 'engine', his soul's earthly residence, is not working properly.

The brain is also divided by medical writers into three segments: imagination, reason, and memory:

> Of these, reason is the principall and chiefe: the other two, because they are her ordinary handmaides, (the one to report; the other, to register and write downe, doe enioy the priuiledges of renowmed excellencie, doe lodge within her royall pallace, and that very neere her own person, the one in her vtter, and the other in her inner chamber.[29]

It is the brain that is the 'royall pallace' or seat of the rational soul. It is divided into three chambers, or ventricles. Reason is clearly located on the middle cell, the memory in the hindmost cell and imagination and common sense at the front. In *Hamlet*, Shakespeare repeatedly reminds us of the sovereignty of reason, but imagination and memory also play a significant part in the opening scenes of the play. When Marcellus and Barnardo bring the appearance of the ghost to Horatio's attention, he says 'A mote it is to trouble the mind's eye.' (I. i. 115) Here he is particularly speaking of the effect that this has on the imagination, 'the mind's eye', which he is then at pains to keep in check as he relates the apocryphal account of the walking dead and lunar eclipse in the city of Rome just prior to the murder of Caesar. (I. i. 116–23) When Hamlet pursues the ghost on the battlements Horatio becomes concerned that Hamlet's 'mind's eye' has been troubled even further, 'He waxes desperate with imagination.' (I. v. 87) Imagination, allowed to wander without the controlling force of reason, is madness.

course raising the doubt in Denmark before the ninth century, when there should certainly be no doubt. The anachronism, if there is one in this allusion, is a fitting one. **28** That he adds the limiting factor of his earthly life to the limitless 'Thine evermore' is further evidence of his temporal confusion. Though this paradox and the stilted and clichéd nature of the composition as a whole suggests that it may not be the sincerest of gestures, and that there is method in Hamlet's madness. **29** Laurentius, pp 73–4.

His imagination left thus to run amok on the image of his murdered father's ghost, Hamlet is bid not to reason, but to remember. 'Remember thee? / Ay, thou poor ghost, whiles memory holds a seat / In this distracted globe.' The multiple resonances of the word 'globe' are worth elucidating. The world is distracted, that is, mad; the audience in the Globe theatre are distracted, that is entertained by the spectacle of the ghost and the play; and most significantly in terms of medical theory while he retains the faculty of memory – which resides in his globular head – he promises to remember. However, he has grounds for fearing that he may not retain it long having waxed desperate 'with imagination' at the sight of the ghost who bids him 'remember'.

The excellence of man

Many of the medical treatise writers introduce their material by commending mankind and the excellence of his body and mind. Bright has a few phrases in his dedicatory epistle that are of particular significance. The first concerns the excellence of the mind, 'Of all other practise of phisick, that parte most commendeth the excellency of the noble facultie, [...] the instrument of reason, the braine'; the second, the excellence of man in distinction to animals, 'the excellencie of man appeareth aboue all other creatures'; the third concerns the power of the mind to devise cures for the infirmities of man, 'The dayly experience of phrensies, madnesse, lunasies, and melancholy cured by this heauenly gift of God'; and, finally, a matter to please the learned students of philosophy, 'discourse of reason'.[30]

Laurentius also writes about the 'excellence of man' beginning by citing Abdalas the Saracen who praised man's adaptability and flexibility. Man is, he says, like a Proteus or a chameleon. The wise men of Egypt call him 'a mortall God', and Mercury calls him 'fellow companion with the Spirits aboue.' Pliny, he tells us, calls him 'the ape or puppie of nature, the counterfeit of the whole world, the abridgement of the great world.' Perhaps most significantly of all, he tells us that the 'Diuines' (he does not specify which ones), 'haue called him, euery maner of creature, because he hath intercourse with euery maner of creature; he hath a being, as haue the stones; life, as haue the plants; and sence or feeling as the beasts; and vnderstanding, as haue the Angels.' He concludes:

> In briefe, this is the chiefe and principall of Gods worke, and the most noble of all other creatures. But this his excellencie, whereby he is more glorious then all the rest, is not in respect of his bodie, although the shape thereof bee more exquisite, better tempered, and of more comely proportion then any other thing in the world [...] This noblenes (I say) commeth not of the bodie, which consisteth of matter and is corruptible: [...] It is

30 Bright, 'The Epistle Dedicatorie'. The phrase is also used by Hamlet (I. ii. 150).

> the soule alone whereby he is so renowmed, being a forme altogether celestiall and diuine, not taking his originall from the effectuall working of any matter, as that of plants and beasts doth.[31]

Hamlet has a corresponding speech. He discusses the seeming wonder of creation, but draws on the religious tradition of the contrariety of the composite of the corrupt flesh of the body and purity of the aspiring soul:

> What piece of work is a man! How noble in reason, how infinite in faculty, in form and moving how express and admirable, in action how like an angel, in apprehension how like a god – the beauty of the world, the paragon of animals! And yet to me what is this quintessence of dust?
>
> (II. ii. 303–8)

Hamlet engages directly with the same material as the medical treatise writers. Particularly clear is the debt to Bright. His 'how noble in reason' surely owes something to Bright's 'Of all other practise of phisick, that parte most commendeth the excellency of the *noble* facultie, [...] the instrument of *reason*, the braine'. (Emphasis added) Hamlet's 'paragon of animals' on its own would not suggest a reference to Bright's description of 'the excellencie of man appeareth aboue all other creatures,' though it expresses the same idea. Hamlet also uses Bright's phrase, 'discourse of reason', in his earliest soliloquy (I. ii. 150). Bright's use of the word 'sovereignty' in conjunction with 'reason' in this passage may also lend something to Horatio's warning to Hamlet not to follow the ghost lest it assume 'some other horrible form / Which might deprive your sovereignty of reason / And draw you into madness' (I. iv. 72–4). Ophelia also links the two words when she laments that she must see in him 'that noble and most sovereign reason / Like sweet bells jangled out of tune and harsh…' (III. i. 159–60). These instances may also recall Bright's insistence that all the parts and spirits of the body be ruled by 'the mind's commandment', that ordinarily nature 'commandeth only by one sovereignty: the rest being vassals at the beck of the sovereign commander'.[32] Laurentius' 'Diuines', who contrast man with plants, beasts, and angels, surely lend something to Hamlet's 'man', 'how noble in reason, how infinite in faculties, in form and moving how express and admirable, in action how like an angel, in apprehension how like a god' (II. ii. 304–5). And perhaps, as in the final part 'apprehension [...] like a god', he also draws from Laurentius in the lines, 'Sure he that made us with such large discourse, / Looking before and after, gave us not / That capability and *godlike reason* / To fust in us unus'd.' (IV. iv. 36–9; emphasis added)

31 Laurentius, pp 72–3. **32** Bright, p. 61.

Thomas Wright's *The passions of the mind in general* does not, like the others, commend in its introduction 'the excellencie of man'. Instead Wright chooses to recommend an intimate knowledge of the passions in order to avoid the sins that taint God's creation. Each nation he has seen tends towards one particular vice; 'for as I have seen by experience,' he tells us, 'there is no Nation in Europe that hath not some extraordinary affection either in pride, anger, lust, inconstancy, gluttony, drunkenness, sloth, or such like passion.' Wright tells us of the dangers of passions: 'I myself have seen some, Gentlemen by blood and Noblemen by birth, yet so appassionate in affections that their company was to most men intolerable [...]'[33] He promotes the understanding of passions saying, 'in fine, every man may by this come to a knowledge of himself, which ought to be preferred before all treasures and riches.'[34] Significantly, his list of excessive passions to be avoided is almost a list of the seven deadly sins. Envy and avarice are missing from the list and have been replaced by inconstancy and drunkenness, two sins that have particular relevance for Hamlet. His mother has been guilty of inconstancy to his father, and the king's 'rouse' (I. iv. 8) causes him to expand upon the particular fault of the Danes – that they are 'drunkards' (I. iv. 19). Wright's association of excessive passions with sin illustrates very plainly the opposition at the heart of much of the rhetoric of the medical treatises and indeed one of the key problems that Hamlet faces. Reason is the gift from God that comes from the soul and is granted to humans alone of the animals on earth; passions come from the physical part of our binary composition, the earthly, corrupt flesh that will eventually return to the dust from whence it came. As such, our passions must, in a sane and godly man, be over-ruled by reason. It is because of this that Hamlet so admires Horatio.

Passion's slave

Hamlet sees Horatio as the perfectly balanced man. As he says:

> Horatio, thou art e'en as just a man
> As e'er my conversation cop'd withal.
> [..............................]
> ... and blest are those
> Whose blood and judgement are so well commeddled
> That they are not a pipe for Fortune's finger
> To sound what stop she please. Give me that man
> That is not passion's slave, and I will wear him
> In my heart's core [...] (III. ii. 54–5; 68–73)

33 Thomas Wright, *The passions of the mind in general* [1601] ed. William Webster Newbold (London, 1986), p. 92. 34 Ibid., p. 89. Polonius ends his advice to Laertes similarly, 'This above all: to thine own self be true'. (I. iii. 78)

Hamlet's description of Horatio combines the notion that balance is the perfect state of health in the humoural system with the ideal that reason should govern emotions. Bright tells us how a 'just proportion' in the 'mixture of the elements' or the 'humours' of the body 'breedeth an indifferency to all passions'.[35] Hamlet means in part to compliment his friend's balanced humoural system in contrast to his own disposition. But the balance between 'blood and judgement' also refers more particularly to the proper relationship between the earthly part of Horatio's composition, 'blood', and his God-given reason, 'judgement'. Hamlet is not balanced as Horatio is.

Wright finds that the evidence of an extreme passion which overrules is recognizable in the face of a man, 'for as a Rat running behind a painted cloth betrayeth herself, even so a Passion, lurking in the heart, by thought and speech discovereth itself'.[36] Polonius' choice of hiding place is unfortunate to say the least, considering the passions evident in Hamlet's chastisement of his mother in the 'closet scene.' (III. iv) When Hamlet stabs through the arras, he exclaims, 'How now? A rat!' (III. iv. 22) If nowhere else in the play, Hamlet, at least at this moment, has allowed his passions to overmaster his reason.

II: CURES AND TREATMENTS

Travel

Claudius' response to the murder of Polonius is not to punish the crime openly, but rather to proceed with his plan to send Hamlet to England, using a recognized treatment for madness as a cover for the dispatch of his nemesis:

> Haply the seas and countries different,
> With variable objects, shall expel
> This something settled matter in his heart,
> Whereon his brains still beating puts him thus
> From fashion of himself. (III. ii. 173)

When the Grave-digger is asked why Hamlet was sent into England, he replies, 'Why, because a was mad. A shall recover his wits there.' (V. i. 146–7) This clearly is the credible reason given out to the common multitude by Claudius. It is also, Laurentius explains, especially used for the treatment of distraction caused by love.

> Remouing, that is to say, the chaunging of the ayre, is one of the rarest remedies, because that vnder colour of that wee may bestow him in some

35 Bright, p. 97. 36 Wright, p. 148.

> remote place, and send him quite out of the countrie: for the sight of his mistresse doth daily blow vp the coles of his desire, and the only reciting of her name serueth as a baite for his ardent affections to bite vpon.[37]

Sex

Another treatment, solely for those suffering from love, stands in complete contrast to this. As opposed to sending the patient away from the object of his affection, whose presence stirs up the heat of his passion, the patient may be treated by allowing him to fulfil his sexual urges. Drawing on the writings of Arnaldus de Villanova, Carol Falvo Heffernan explains what is happening in the body of the lover:

> The heat of *amor hereos* – closely associated with mania – is caused by the overheating of the vital spirit by a pleasing form. This vital spirit, in turn, generates heat in the animal spirit, which inflames the middle ventricle of the brain, the seat of the faculty of estimation.[38]

The fulfilling of the sexual desires of the patient may in part relieve this heat. It was believed that sperm or 'seed' is blood heated so much that it turns white. For this reason Lawrence Babb tells us: 'When blood abounds in a body, seed also abounds. Abundance of seed causes amorous disposition.'[39] The release of the heat of erotomania, that is, love madness, is best achieved by the satiation of the desire. Laurentius speaks specifically of a male patient in this regard, but the principle is the same for the female case: 'it is certaine that the principal cause of the disease which is this burning desire, being taken away, the diseased partie will finde himselfe marueilously relieued […]'[40] If this is retained, as in the case of unrequited love, it conflates the issue of the overheated spirits. It is important of course that the honour of the parties involved be protected, and so the solution in this case would preferably be their marriage. While Laurentius gives the physician his due mention, he recommends coitus first: 'There are two waies to cure this amourous melancholie: the one is, the inioying of the thing beloued: the other resteth in the skill and paines of a good Phisition.'[41] There may be a subtle allusion to this remedy in Gertrude's encouragement of Ophelia:

> And for your part, Ophelia, I do wish
> That your good beauties be the happy cause
> Of Hamlet's wildness; so shall I hope your virtues
> Will bring him to his wonted way again,
> To both your honours. (III. i. 38–42)

37 Laurentius, p. 123. **38** Carol Falvo Heffernan, *The melancholy muse: Chaucer, Shakespeare and early medicine* (Pittsburgh, 1995), p. 72. **39** Lawrence Babb, *The Elizabethan malady: a study of melancholia in English literature from 1580 to 1642* (1951; East Lansing, MI, 1965), p. 129. **40** Laurentius, p. 121. **41** Ibid.

Gertrude is hoping for a marriage, a match that will both satisfy the young couple's romantic desires and cure Hamlet's madness.

Entertainment, honest mirth and good company

Another recommendation that the medical treatise writers make for more general melancholy is company and companionship. Boorde offers the opinion that 'There is nothing that doth comfort the heart so much, beside God, as honest mirth and good company.'[42] Thomas Wright also affirms this,

> It is good also to have a wise and discreet friend to admonish us of our Passions when we err from the path and plain way of Virtue; for, as I have often said, self-love blindeth much a man, and another may better judge of our actions than we can ourselves [...] if a man might have such a friend I would think he had no small treasure.[43]

This is Claudius' apparent reason for inviting Rosencrantz and Guildenstern to Denmark:

> I entreat you both
> That, being of so young days brought up with him,
> And since so neighboured to his youth and humour,
> That you vouchsafe your rest here in the court
> Some little time, so by your companies
> To draw him on to pleasures, and to gather,
> So much as from occasions you may glean,
> Whether aught to us unknown afflicts him thus
> That, opened, lies within our remedy. (II. ii. 10–18)

Though, as always, his intentions are to be viewed with suspicion, his declared intention is medically feasible. Guildenstern reflects this belief when he responds, 'Heavens make our presence and our practices / Pleasant and helpful to him.' (II. ii. 38–9) Hamlet shows himself to be aware of the benefit of a well-balanced and honest confidant when he takes Horatio into his confidence. Hamlet's praise of Horatio's balanced and reasonable nature (III. ii. 54–73), as discussed above, is of the utmost importance in this regard. If Hamlet cannot rely on himself to be governed by reason, his friend and companion must be.

Laurentius also recommends constant activity and entertainment, again particularly to those stricken by love,

42 Boorde, p. 17. **43** Wright, p. 148.

> It will bee good for him, to lodge in the fields, or in some pleasant house; to cause him to walke often; to keepe him occupied euery houre with one or other pleasant pastime; to bring into his minde a hundred and a hundred sundrie things, to the end he may haue no leisure to think of his loue; to carrie him out a hunting; to the fenceschoole; to holde him vp sometimes with fine and graue stories; sometime with pleasant tales; and therewith to haue merrie musicke [...][44]

In John Ford's *The Lover's Melancholy* (1628), the melancholy Prince Palador is prescribed the entertainment of a play as an attempted cure.[45] In *The Antipodes* (1640) by Richard Brome, the character of Joyless is tricked into watching a play as a way of curing his mad jealousy, while performing in it cures his son's travel mania.[46]

The arrival of the players and their performance at court in *Hamlet* may then be seen as a form of treatment for Hamlet. His mother Gertrude is anxious to know whether Rosencrantz and Guildenstern have assayed him to any pastime, to which they enthusiastically relate the arrival of the players and Hamlet's intent to have them perform. Indeed, this is greeted by Claudius as though it were the arrival of a famous doctor rather than a company of players:

> [...] it doth much content me
> To hear him so inclin'd. Good gentlemen,
> Give him a further edge, and drive his purpose on
> To these delights. (III. i. 24–7)

The 'dark room and the whip'

One other treatment for people suffering from madness, though it is as much for the protection of others as it is for the treatment of patients, is the practice of locking them up. During the Renaissance period this practice seems to have been believed to be the most prudent treatment for lunatics. The practice of locking up madmen, particularly in a dark room, is alluded to in *Twelfth Night*, when Sir Toby decides what should be done with Malvolio, 'Come, we'll have him in a dark room and bound.' (III. iv. 136–7) Dr Pinch, the quack of *The Comedy of Errors*, also recommends this procedure for Antipholus and Dromio, 'both man and master is possess'd, / I know it by their pale and deadly looks; / They must be bound and laid in some dark room.' (IV. iv. 90–2) Andrew Boorde writes,

> There is no man the which have any of the kinds of madness but they ought to be kept in safeguard, for divers inconvenience that may fall, [...]

44 Laurentius, p. 123. 45 John Ford, *The Lover's Melancholy* in Marion Lomax (ed.), *'Tis Pity She's a Whore and other plays* (Oxford, 1995) III. ii. 46 Richard Brome, *The Antipodes* in Anthony Parr (ed.), *Three Renaissance travel plays* (Manchester, 1999).

> I do advertise every man the which is mad, or lunatic, or frantic or demoniac, to be kept in safe guard in some close house or chamber, where there is little light [...] And see that the mad man have no girdle, except it be a weak list of cloth (to prevent) hurting or killing him self.[47]

Curiously, Boorde mentions the dangers of wall hangings to mad men, 'Also the chamber or the house that the mad man is in, let there be no painted cloths, nor painted walls, nor pictures of man or woman, or fowl or beast; for such things maketh them full of fantasies.'[48] Again, Polonius' choice of hiding place in the 'closet scene' is unfortunate. (III. iv.)

Purgation and blood letting

Release of corrupted, undesirable, or diseased fluids from the body is often achieved by bloodletting, or purgatives. Bright recommends the practice of blood letting with particular reference to the melancholic humour, 'The humour requireth euacuation, and emptying: and because your body is not only melancholicke vnder the ribbes but the whole masse of your blood is chaunged therewith: it shall be first necessarye to open a vaine: [...]'[49] This is recommended, albeit in jest, by Hamlet as a treatment for Claudius' 'choler' after the play scene. (III. ii. 291–9) Claudius is of a similar mind, when he sends his letters to England demanding the 'present death of Hamlet [...] / For like the hectic in my blood he rages, / And thou must cure me.' (IV. iii. 68–70)

III

Hamlet's description of his body as a 'machine', and his assertion that he cannot 'count' suggest a connection with his earlier epithet, 'The time is out of joint'. (I. v. 196) His body, like Bright's clock, tainted by melancholy, is unable to keep time. The age in which he lives is 'out of joint'. There is somehow a correspondence between the disjunctions in the Danish state and the distemper of Hamlet's mind. Laertes gives a possible explanation to this in his warning to Ophelia to be wary of Hamlet's courtship:

> He may not, as unvalu'd persons do,
> Carve for himself, for on his choice depends
> The sanity and health of this whole state;
> And therefore must his choice be circumscrib'd
> Unto the voice and yielding of that body
> Whereof he is the head. (I. iii. 19–24)

47 Boorde, pp 51–2. **48** Ibid., p. 52. **49** Bright, p. 269.

What Laertes refers to is the concept of the 'body politic', whereby the ruler of a state is accounted the metaphorical 'head' of the body of the realm. Hamlet's 'greatness' as the heir elect, or future 'head' of the state of Denmark makes 'his choice' of importance not only to himself but to the whole 'body' of the state. However, what Laertes suggests (that the 'sanity and health of this whole state' can be affected by Hamlet's own sanity and health) seems to fly in the face of what Edmund Plowden writes in his *Commentaries, or Reports* (1578) – reports from the law courts of Elizabeth I's reign – in which he explains some of the elements of the theory of the 'King's two bodies':

> For the King has in him two Bodies, *viz.*, a Body natural, and a Body politic. His Body natural (if it be considered in itself) is a Body mortal, subject to all Infirmities that come by Nature or Accident, to the Imbecility of Infancy or old Age, and to the like Defects that happen to the natural Bodies of other People. But his Body politic is a Body that cannot be seen or handled, consisting of Policy and Government, and constituted for the Direction of the People, and the Management of the public-weal, and this Body is utterly void of Infancy, and old Age, and other natural Defects and Imbecilities, which the Body natural is subject to, and for this Cause, what the King does in his Body politic, cannot be invalidated or frustrated by any Disability in his natural Body.[50]

The first court scene in *Hamlet* is filled with the rhetoric of the body politic. It also however gives the impression of a state body under threat and in the grips of emotional conflict. Claudius and the court have been lamenting King Hamlet's death with 'hearts in grief, and our whole kingdom / […] contracted in one brow of woe'. (I. ii. 3–4) Now he plans also to celebrate his marriage, 'With an auspicious and a dropping eye'. (I. ii. 11) The initial image is of a face contorted with contrary emotions. The first matter of court business involves the dangers posed by military posturing of ambitious Fortinbras. 'Holding a weak supposal of our [Claudius'] worth, / Or thinking by our late dear brother's death / Our state to be disjoint and out of frame …' Fortinbras has thought to capitalize on a perceived weakness in both Claudius and, by extension, in the Danish state. He is capable of this martial threat it seems because, 'Norway […] / […] impotent and bedrid, scarcely hears / Of his nephew's purpose […]' (I. ii. 28–30) Though it is the king of Norway that Caudius refers to, he does so in terms that agree with the idea of the 'body politic' under which the king is the personification of Norway in being its head of state. The ambassadors again refer to his 'sickness, age, and impotence'

50 Edmund Plowden, *The Commentaries, or Reports of Edmund Plowden*, 2 vols (rpt. London, 1816) i, 212a; cit. Ernst H. Kantrowicz, *The king's two bodies: a study in medieval political theology* (1957; rpt. Surrey, 1981), p. 7.

(II. ii. 66) when they return. By the rhetorical suggestion then, not only is the king of Norway sick but so is Norway itself. By the same rhetorical device, when Fortinbras perceives a weakness in Claudius, he also perceives a 'weakness' in Denmark that corresponds to its being 'disjoint and out of frame.' The resonance with Hamlet's the 'time is out of joint' is palpable. From this perilous beginning the play continues to refer to the body politic, and in terms that see this disjunction worsen rather than improve.

We have already noted Laertes' concern for the 'sanity and health of this whole state' (I. iii. 21) In the scene that follows Marcellus proclaims that there is 'Something is rotten in the state of Denmark.' (I. iv. 90) When Claudius reveals his intention to have Hamlet dispatched to England, Guildenstern's reply is a rather cannibalistic reassertion of the concept of the body politic, 'Most holy and religious fear it is / To keep those many bodies safe / That live and feed upon your Majesty.' (III. iii. 7–9) But as Rosencrantz continues the attempt to justify Claudius' actions, the imagery he chooses offers both a reminder of the early appearances of the Ghost king, and a warning of what is to come,

> The single and peculiar life is bound
> With all the strength and armour of the mind
> To keep itself from noyance; but much more
> That spirit upon whose weal depends and rests
> The lives of many. The cease of majesty
> Dies not alone, but like a gulf doth draw
> What's near it with it. It is a massy wheel
> Fixed on the summit of the highest mount,
> To whose huge spokes ten thousand lesser things
> Are mortised and adjoined, which when it falls
> Each small annexment, petty consequence,
> Attends the boist'rous ruin. Never alone
> Did the King sigh, but with a general groan. (III. iii. 10–23)

With his usual lack of perception Rosencrantz envisages the decline and death of the head of state bringing the whole state down with him. King Hamlet has already died and indeed the corruption that started with the initial death of that 'head' of the body politic has spread to the rest of the state, and its demise is following just as Rosencrantz imagines.

For his part, Claudius imagines Hamlet to be the infectious element within the body of the state and has called on the metaphorical medical assistance of England, 'Do it, England; / For like the hectic in my blood he rages, / And thou must cure me.' (IV. iii. 68–70) The infection, however, had begun with Claudius' initial introduction of poison to the system. He poured poison in the ear of the

Danish King and although the body politic survived the death of that head of state, the metaphorical poison remained within the body of the Danish court infecting all within it. There are two families in the play that represent the body of the state. As Claudius says at the beginning of the play, 'The head is not more native to the heart, / The hand more instrumental to the mouth, / Than is the throne of Denmark to [Polonius].' (I. ii. 47–9) Both families become infected by this poison. Hamlet is initially the part of the body most profoundly affected but the attempts to treat this melancholy aspect of the court body prove futile because the disease is one stemming from a guilty conscience, of which Bright writes:

> Here no medicine, no purgation, no cordiall, no tryacle or balme are able to assure the afflicted soule and trembling heart, now painting vnder the terrors of God: [...] In this affliction, the peril is not of body, and corporall actions, or decay of servile, and temporall vses, but of the whole nature soule and body cut of[f] from the life of God, and from the sweet influence of his fauour, the fountaine of all happiness and eternall felicity.[51]

The infection spreads like the plague. Polonius dies as a result of Hamlet's distemper. Ophelia goes mad as a result of the combined affect of her father's death and Hamlet's rejection. This in turn causes a rebellion that is only temporarily subdued. Finally, the play closes, appropriately as it began with a 'leperous distilment' coursing through the 'natural gates and alleys' of the remaining parts of the body of the state, (I. v. 64–7) Laertes, Gertrude, Claudius, and Hamlet. Though the play expresses a range of contemporary medical theories and the members of the court seek to implement a number of legitimate cures, there is no medicine that can cure Hamlet or Denmark, and it falls to the Norwegian hands of Fortinbras to care for the body and to Horatio to remember. On one level the representation of the Danish court echoes, in the course of the play, the invocation of the ghost. 'Remember me', (I. v. 91) had been the Ghost's command. From that point on, like a headless spectre stumbling through its final apparition, the Danish state struts and frets its final hours upon the stage, the corruption of that fatal wound to the head spreading through the rest of the body politic. At last, a dying Hamlet, the final member of the Danish body politic, bids Horatio, 'tell my story'. (V. ii. 354)

51 Bright, pp 189–90.

The hippopotamus prince; or, the perils of visiting England in 1850

KATE HEBBLETHWAITE

Although an unlikely candidate for the exposition of a seachange in biological thought, the figure of the hippopotamus was a fundamental participant in the evolutionary debates of the mid-nineteenth century. Utilized more often than not for its comic potential, this hulking yardstick is an excellent indicator of the groundswell of opinion that followed in the wake of more weighty scientific conjecture. Heavily influenced by the contemporary *zeitgeist*, the cultural triggers of subject-specific comedy not only point toward pertinent trends in social thought, but are also indicators of that fine line between acceptance and unease where laughter serves a two-tier function of inclusion and exclusion. Those who are 'in' on the joke, in other words, being thus proven as in some way superior to those against whom such humour is directed. As this essay will argue, in the thirteen years spanning the arrival of Obaysach at London Zoo in 1850 to the publication of Charles Kingsley's *The Water Babies* in 1863, the fulsome figure of the hippo served a hitherto unacknowledged function as an important gauge for shifting attitudes about evolution theory.

While the figure of the hippopotamus is the catalyst for many of the arguments made in this essay, the habitual vehicle for the dissemination of such scientific theories to a non-scientific audience throughout the nineteenth century was the popular press and, indeed, literature itself. At every turn, authors responded to advances in scientific understanding about the world and humanity's place within it. Writers that would now be termed 'creative' readily explored the implications of scientific theories while scientists themselves quoted well-known poets in their textbooks. Indeed, the notion of a definitive schism between the language of literature and the language of science only really became apparent toward the end of the nineteenth century.[1] Charles Lyell's *Principles of Geology* (1830–3), for

1 For an excellent overview of the compelling links between the two genres, see Laura Otis (ed.), *Literature and science in the nineteenth century: an anthology* (Oxford, 2002).

example, couched his arguments for a uniformitarianist approach to geology within literary references to Milton, Scott and Wordsworth, while the oft-quoted 'entangled bank' metaphor with which Charles Darwin concluded *The Origin of Species* (1859) eased his readers' understanding of the complexities of the interconnected nature of biological organisms. In turn, the theories of Darwin and the consequences evolution theory augured for humankind's perception of itself prompted a wealth of considered, reflective or anxiety-ridden rejoinders by writers of fiction. From George Eliot's contemplative ideas on social as well as biological evolution in *Middlemarch* (1871–2) to H.G. Wells' shocking vision of undirected human development in *The Time Machine* (1895), evolution theory impelled authors to explore the fundamentals of human experience above and beyond the dictates of strictly fictional parameters.

In any discussion of the theory of evolution and its impact on nineteenth-century thought, however, it is worth remembering that the dethroning of man from his self-appointed position as the pinnacle of biological creation itself did not derive solely from the publication of Charles Darwin's *The Origin of Species*. Indeed this work neither made specific mention of the applications of its theories to the evolution of man, nor were the ideas about natural selection and competitive survival themselves wholly revolutionary. As James Turner has appositely argued, 'If the walls came tumbling down at the sound of Darwin's trumpet, it was because the humble termites had nibbled for years.'[2] Challenging the rigidity of the Great Chain of Being[3] with the notion that species were not specially and distinctly created, but that they had evolved from earlier forms, had been variously attempted in the century preceding that of Darwin's own works. Carl Linnaeus (1707–78), George-Louis Leclerc, Comte de Buffon (1707–88), Erasmus Darwin (1731–1802) and Jean–Baptiste de Lamarck (1744–1829) are but four names which deserve due attention for their work that, variously, enabled nineteenth-century biologists and geologists to formulate claims for the gradual, millennial processes of biological advancement.

In the nineteenth century itself, advances in the geological and palaeontological sciences prompted Lyell (1797–1875) to argue in *Principles of Geology* for a concept of a geological time scale far greater than hitherto supposed. In 1844, the evolution debate took another step forward with Robert Chambers' *Vestiges of the Natural History of Creation* which, although scientifically inaccurate in much of its detail, brought together geology and biology in its argument for the organic development and the origin of species by natural descent. In turn, both Chambers

2 James Turner, *Reckoning with the beast: animals, pain and humanity in the Victorian mind* (Baltimore and London, 1980), p. 61. 3 The theory of the Great Chain of Being had been a prevalent belief throughout the Middle Ages and the Renaissance. It established the fixed structure and closed hierarchy of all creation arguing that all forms of life had been created with the creation of the universe and that as such there was no room for any new life-forms to appear.

and Lyell's works inspired Alfred Russel Wallace (1825–1913), who combined the latter's view of geological causes with his own directional view of the fossil record to refine his examination of the geographical distribution of species. In his paper 'On the Law which had regulated the introduction of new species' (1855), Wallace compared the evidence of geography and geology to argue that: 'Every species has come into existence coincident both in time and space with a pre-existing closely allied species.'[4] Wallace's Law obviated the Hand of God, the living world being, 'clearly derived by a natural process of gradual extinction and creation of species from that of the latest geological periods.'[5] Although not directly proclaiming it, the natural process Wallace inferred was evolution. Spurred into action by Wallace's further refinement of the ideas outlined in his paper, 'On the tendency of varieties to depart indefinitely from the original type' (1858), Darwin published his own theory of evolutionary change and adaptation the following year. The widespread attention subsequently garnered by Darwin's works facilitated a proliferation of evolutionary theorists in the later decades of the nineteenth century, most notable among whom included Thomas Henry Huxley (1825–95), William Bateson (1861–1926), Francis Galton (1822–1911), Ernst Haeckel (1834–1919), and August Weismann (1834–1914).

This brief run-through of evolution's 'Who's Who' is by no means exhaustive.[6] Its aim, however, is to illustrate the idea that the theory of evolution was itself a gradually evolving concept, neither starting nor ending with Charles Darwin. This is particularly important to bear in mind in relation to developing ideas about man's relationship with the animal kingdom. For centuries, Western theologians and philosophers had accepted the Platonic duality of animal and human souls, and science offered no evidence that undermined this concept. These unambiguous boundaries between man and animal were upheld by the longstanding conviction that what defined humanity was the concept of reason. Humans possessed it, animals did not. If a man yielded to his baser emotions or desires he was essentially surrendering his reason, his claim to call himself human, and he would become instead animal-like. Maintaining moral propriety, however, his 'humanity' was safe. The popular morality play, *Everyman* (*c.*1509–19), for example, warned that if mankind lived solely for pleasure and the gratification of his desires then: 'In their life and wicked tempests, / Verily they will become much worse than beasts.'[7] Despite plumbing moral bestial depths through the capitulation to temptation or folly, however, the human form

4 Alfred Russel Wallace, 'On the Law which had regulated the introduction of new species' in Andrew Berry (ed.), *Infinite tropics: an Alfred Russel Wallace anthology* (London, 2002), p. 49. 5 Ibid., p. 37. 6 For a more comprehensive review of the history of evolution that I am able to give here, see Philip Whitfield, *The natural history of evolution* (London, 1993); Carl Zimmer, *Evolution: the triumph of an idea* (London, 2002); Peter J. Bower, *Evolution: the history of an idea* (Berkeley, CA, 2003). 7 *Everyman* in A.C. Cawley (ed.), *Everyman and medieval miracle plays* (London, 1993), p. 200.

retained its autonomous physiognomy: the eternal consolation of Divine grace and favour.

In the nineteenth century, however, as the biological sciences unremittingly revealed humans and animals to be composed of effectively the same fundamental biological elements, inflexible notions of human rationality and bestial animality blurred, with the frightening potential for an osmotic intermingling of form and hierarchy. Notions of fluidity of identity and evolutionary development raised the possibility of animals developing reason, while the apparently accidental nature of mankind's dominion of the Earth, and his essential inconsequentiality in the vast time-frame of created life-forms, gave rise to fears of the prospect of his own concomitant deterioration. The demotion of humankind in the nineteenth century was a process achieved across a wide spectrum of social and scientific disciplines. Humanity's lowly, non-divine origins were traced by anatomists and zoologists who dissected to find internal similarities; by the embryologist who posited that ontogeny recapitulated phylogeny;[8] and by the biologist who, with the aid of his microscope, revealed the fundamental parallels between the cellular composition of humans and animals. In his monumental work on the anatomy of the human body, *Anatomy, descriptive and surgical* (1858), Henry Gray revealed man in all his physicality, describing the brain, not as a God-given marvel of rationality and reason, but as 'the largest portion of the encephalon'.[9] No allowance was given over to a spiritual or moral organism; all was subsumed to the corporeal and biological. Six years later, an article in the *Anthropological Review* conceded that:

> Man has looked down at the animals only through the deceitful prism of his own pride, and his own unreasoning individuality [...] Man does not form an order apart from the rest of the animal world; he is linked to that world by humiliating but indissoluble ties of resemblance and connection.[10]

Evolutionary theory, however, was perhaps the most emotive and destructive of those forces which sought to topple mankind from his self-appointed position

8 Ernst Haeckel's Biogenetic Law (1866) explained how ontogeny (individual development from foetus to adult) recapitulated in abbreviated form the phylogenetic evolution of the human species. In other words, 'an organism, during the course of its embryonic growth, passes through a series of stages representing adult ancestors in their proper historical order.' See Stephen Jay Gould, 'Freud's evolutionary fantasy' in *I have landed: splashes and reflections in natural history* (London, 2003), p. 148. The idea was not necessarily a new one, however. Robert Chambers' *Vestiges* had posited a similar theory, arguing that, 'each animal passes, in the course of its germinal history, through a series of changes resembling the *permanent forms* of the various order of animals inferior to it in the scale.' Chambers also applied his idea to *Homo sapiens*: 'His organization gradually passes through conditions generally resembling a fish, a reptile, a bird, and the lower mammalian, before it attains its specific maturity.' See Chambers, *Vestiges of the natural history of creation* (New York, 1969), pp 198, 199. **9** Henry Gray, *Anatomy, descriptive and surgical* (Bath, 2002), p. 468. **10** Philalethes, 'The distinction between man and animals', *Anthropological Review* 2 (August 1864), 157, 163.

as a divinely created central figure in the universe. While Darwin's *The Origin of Species* certainly emphasized the close relation of species and conceived all living forms as having their source in a common ancestry, only a single, enigmatic, reference was made to the possible application of the theories to *Homo sapiens* – 'Light will be thrown on the origin of man and his history.'[11] The parallel relationship between man and beast (and the inferential downgrading of man from divinely blessed to bestially equivalent) could be, and was, nevertheless readily surmised. The extent of mankind's relegation was categorically stated in Darwin's second work on evolution, *The Descent of Man* (1871). This dealt specifically with the evolution and competitive survival of *Homo sapiens*, notoriously putting apes in the human family tree and making the races one family, diversified by sexual selection. Concentrating itself specifically with man's origin and development, *The Descent* worked to challenge the self-styled superiority of humankind: 'It is only our natural prejudice, and that arrogance which made our forefathers declare that they were descended from demi-gods, which leads us to demur to this conclusion.'[12] To Darwin, it was no scientific explanation to say that the physiognomical and anatomical similarities between man and ape were as a result of a harmonious design in the mind of the Creator. They could be explained much more credibly by assuming common descent with selective modification, 'the difference in mind between man and the higher animals, great as it is, certainly is one of degree and not of kind [... M]an in all parts of his organization differs less from the higher apes, than these do from the lower members of the same group.'[13]

In arguing that 'man is descended from some lower form',[14] *The Descent of Man* boldly negated other contemporary theories about humanity's origin and subsequent evolution. In particular, Darwin directly challenged the theory of polygenism which held that the different races of man were, in fact, different species:

> If the races of man had descended, as is supposed by some naturalists, from two or more species, which differed from each other as much, or nearly as much, as does the orang from the gorilla, it can hardly be doubted that marked differences in the structure of certain bones would still be discoverable in man as he now exists.[15]

Pointing to numerous physiological and dispositional similarities between various races, and that such 'close agreement' in animal types would lead to the inferential conclusion 'that they are descended from a common progenitor who was thus

11 Charles Darwin, *The Origin of Species*, ed. J.W. Burrow (London, 1985), p. 458. **12** Charles Darwin, *The Descent of Man*, ed. James Moore and Adrian Desmond (London, 2004), p. 43. **13** Ibid., p. 176. **14** Ibid., p. 172. **15** Ibid., p. 207.

endowed', Darwin maintained that '[t]he same argument may be applied with much force to the races of man.'[16] As such, Darwin declared that, while 'American aborigines, Negroes and Europeans are as different from each other as any three races that can be named,' there existed 'many little traits of character, shewing [*sic*] how similar their minds were to ours.'[17] For Darwin, *Homo sapiens* was an homogeneous species, and he was content to acknowledge kinship with all men.

Twenty years before this refutation of polygenism by Darwin, the happy acceptance of the 'native' as the biological equivalent of the 'civilized' Westerner was not so readily undertaken. While, as mentioned above, the evolution of mankind was an area of active investigation in the early decades of the nineteenth century, the egalitarianism of Darwin's viewpoint was relatively rare. In his excellent study of biological determinism, *The mismeasure of man*, Stephen Jay Gould argues that during this pre-Darwinian period 'leaders and intellectuals did not doubt the propriety of racial ranking',[18] numerous methods being devised for the preservation of such gradation. Polygenism itself was by no means a new theory in the nineteenth century: earlier supporters included the eighteenth-century philosophers Voltaire and David Hume. In 1753, for example, Hume declared of the variances between nationalities that, 'Such a uniform and constant difference could not happen [...] if nature had not made an original distinction between these breeds of men.'[19] Likewise, in his notorious *The history of Jamaica* (1774), Edward Long contended in favour of what he described as:

> [the] gradations of the intellectual faculty, from the first rudiments perceived in the monkey kind, to the more advanced stages of it in ape [...] and the Guiney Negroe; and ascending from the varieties of this class to the lighter casts, until we mark its utmost limit of perfection in the pure White.[20]

Arguing at length for the indigenous African to be classified as a separate species, Long maintained that:

> The measure of the several orders and varieties of these Blacks may be as compleat [*sic*] as that of any other race of mortals; filling up that space, or degree, beyond which they are not destined to pass; and discriminating them from the rest of men, not in *kind*, but in *species*.[21]

16 Ibid., p. 208. **17** Ibid., p. 207. **18** Stephen Jay Gould, *The mismeasure of man* (London, 1997), p. 63. **19** David Hume, 'Of national characteristics' in *Essays: moral, political and literary*, ed. Eugene F. Miller (Indianapolis, 1987), p. 208. **20** Edward Long, *The history of Jamaica, or general survey of the antient and modern state of that island: with reflections on its situations, settlements, inhabitants, climates, products, commerce, laws and government* (London, 1970), ii, pp 374–5. **21** Ibid., p. 375.

As such, the most 'ignoble' native tribe, the Hottentots, were more nearly akin to ape than human, he argued:

> I do not think that an orang-outang [*sic*] husband would be any dishonour to an Hottentot female; for what are these Hottentots? – They are [...] a people certainly very stupid and very brutal. In many respects they are more like beasts than men [...] that the orang-outang and some races of black men are very nearly allied, is, I think, more than probable.[22]

Essentially a justification for the propagation and continuation of slavery, Long's polemical propaganda on behalf of the slave-owning plantocracy of Jamaica invoked a brand of theological evolution – that God created the universe in the form of a long series of gradations with the white man at its zenith – as a means to justify social, and economic, ends.

Advocacy of a social policy, as much as the demonstration of scientific 'fact', underlay many nineteenth-century theories of polygenism. An idea that engendered much support in America, as Gould has noted, 'polygenism was one of the first theories of largely American origin that won the attention and respect of European scientists – so much so that the Europeans referred to polygeny as the "American school" of anthropology.'[23] Considered in retrospect, nineteenth-century race theory in America was little more than a thinly-disguised validation for slavery and, while it is not the intention of this essay to trace the history of American polygeny, a couple of examples are necessary to highlight this close bond between social strategy and scientific theory.

Although actively disclaiming involvement with political matters, Louis Agassiz (1807–73), one of the main proponents of polygenism in America, nevertheless overlay his major statement on the subject, 'The diversity of origin of the human races' (1850), with a moral dimension that provided a justification for slavery in all but name:

> We entertain not the slightest doubt that human affairs with reference to the coloured races would be far more judiciously conducted if, in our intercourse with them, we were guided by a full consciousness of the real difference existing between us and them, and a desire to foster those dispositions that are eminently marked in them, rather than treating them on terms of equality.[24]

Naturally occupying the lowest position in the racial hierarchy, the 'submissive, obsequious, imitative negro',[25] in other words, should seek employment befitting

22 Ibid., pp 364–5. **23** Gould, *The mismeasure of man*, p. 74. **24** Louis Agassiz, 'The diversity of origin of the human races', *Christian Examiner* 49 (1850) 145. **25** Ibid., p. 144.

their biologically predetermined servility. Samuel George Morton (1799–1851) compounded the polygenists' argument of separate and unequal with empirical evidence that, he claimed, demonstrated the innate superiority of the white races. Believing that a racial ranking could be established by brain physiology, Morton measured the cranial capacity of brain cavities from sample racial groups with the *a priori* conclusion that brain size was a sure indication of mental ability. His results, published in *Crania Americana* (1839) and *Crania Aegyptiaca* (1844) rather unsurprisingly concluded Caucasians to have the largest cranial capacity, and Negroes the smallest.[26] Although not specifically used as a wholesale justification for the continued practise of slavery in America, the theories of Agassiz, Morton and the American polygenists certainly did not hinder it. The scientific objectification of the dark races as both the social and biological 'other', itself built upon the practise of observation and examination, validated the notion of a racial pecking order that would only finally be comprehensively challenged almost one hundred years later.

It is with this notion of observation as objectification that the remainder of this essay will now turn. Although regarded as a largely American phenomenon, mid-nineteenth-century racial attitudes in Britain, as principally expounded in the popular press, displayed a degree of inherent prejudice that paralleled the polygenism debates. The spectacle that surrounded the arrival of a hippo and a Hindu man in London, initially remarkable for their intensity, also reveal an entrenched attitude toward the foreign 'other' (be it animal or indigenous native) that may more nearly be couched in terms of polygenist rhetoric.

In 1850, hippomania swept through Britain with the arrival at London Zoo of the first live hippopotamus to be seen in Europe in modern times. Captured at the beginning of August 1849 on the island of Fobaysch, in the White Nile, roughly 2,000 miles north of Cairo, the creature was transported to London in May the following year. Named Obaysach, the hippo immediately became a local celebrity. Also arriving to Britain in May 1850 – and coincidentally on the same ship as the hippo – was a second spectacle for public scrutiny, the retinue of His Highness General Jung Bahadoor Ranajee, Crown Prince of Nepal. On a state visit to Queen Victoria, Prince Bahadoor was, 'the first Hindoo of so high a caste who has ever been presented to Her Majesty.'[27] The concurrent arrival of the Hindu and the hippo in London, and the subsequent reportage each amassed, allows for a fascinating insight into attitudes towards both race and evolution as propounded in the popular press of the mid-nineteenth century.

26 Craniometry flourished in the second half of the nineteenth century under the French physician and anthropologist Paul Broca (1824–80) who argued that human races could be ranked in a linear scale of mental worth – itself measurable by cranial capacity. See Gould, *The mismeasure of man*, pp 105–41. **27** *The Times,* 27 May 1850.

From the arrival of the P&O ship Ripon at Southampton, comparisons between the entourage of the Crown Prince and the hippo were subtly propagated. In an article published on 20 May 1850, *The Times* reported that 'These Hindoos [*sic*] will eat nothing but what is killed according to their own notions, so that a large number of kids [...] had to be provided for them for their passage'. Eight lines later the dietary requirements of the hippopotamus are described in much the same manner: 'The hippopotamus [...] is comparatively small, and lives exclusively on milk, its daily consumption being about 80 pints, for the furnishing of which several cows have to be kept on board.'[28] On 27 May, *The Times* again drew equivalence between the man and the animal, remarking that the opulence of the Prince's entourage 'attracted much attention'[29] in the same article as it described the offloading of the hippopotamus from the Ripon prior to its journey to London Zoo for public display. The relative ease with which the disembarkation of the hippo was achieved is contrasted sharply with that of the Nepali prince who, it was reported, was subject to lengthy delays by the Customs authorities' refusal 'to pass the baggage belonging to the Prince Jung Bahadoor Ranajee without the usual formalities of examination.'[30]

Charles Dickens' family magazine, *Household Words*, disregarded any such oblique comparison of the two visitors, candidly highlighting the spectacle they afforded and the relative consequence of each. Having briefly accounted for the presence of the Nepalese entourage on board the Ripon, the reporter states that: 'These latter personages would have been great objects of attraction under any other circumstances; but what could stand against such a rival as the occupant of the great house and bath on the main deck?'[31] Ostensibly just a throwaway remark, it is nevertheless indicatory of the unconscious ranking of worth along race and species lines that the article advances: the Nepalese royalty are a less interesting spectacle than the novelty of a hippo, and both ultimately serve the purpose of entertaining the curious English crowds.

In many of the articles on the subject of the hippo and the Hindu, not only are equivalences made between the animal and the human, but the former is afforded a greater degree of humanity than its bipedal counterpart. 'The Diary of the Hippopotamus' in *Punch* recounted the daily routine of the animal, reporting that, 'At seven he has his pail of porridge and maize, which he refers to as tea or coffee. After that he washes his hands – we mean his feet – in the tank which is put in his room as his wash-hand basin.'[32] Concern over the potential fatigue of the hippo, resulting from its hectic round of social engagements, further lead *Punch* to suggest that the animal take an invigorating trip to the seaside, as much for the benefits to his social standing as to his health:

28 *The Times*, 20 May 1850. **29** *The Times*, 27 May 1850. **30** Ibid. **31** *Household Words* 19, 3 Aug. 1850, p. 448. **32** *Punch* 19, 6 July 1850, p. 20.

Fig. 1: 'The delicate state of the hippopotamus. It is ordered change of air and a little sea-bathing.' *Punch* 19, 27 July 1850.

> The fashionable lions will soon be 'running down' to the sea-side, and if such refreshment is required for the 'fashionable lions,' why not for that greatest of all the lions of the season, the Hippopotamus? We think it is high time that the poor animal obtained the benefit of the invigorating sea breeze after the labours of the past few months, during which he has been the 'observed of all observers,' and the centre of attraction to the whole metropolis.[33]

The accompanying illustration to this article, depicting the hippo clad in bathing costume and being coaxed into a bathing hut, exemplifies this anthropomorphism.

In contrast, the jocular dehumanization of General Ranajee continued in the press throughout the Nepalese prince's stay. On the preceding page to the proposed hippopotamus holiday, *Punch* reported on the foreign prince's attendance at the Chiswick fête, describing the excitement his presence prompted in terms more appropriate to a safari hunt than a tea party:

33 *Punch* 19, 27 July 1850, p. 50.

> The Nepaulese [*sic*] Ambassador was hunted under a scorching sun, for full two hours, in the most determined manner. How he managed to keep up so long was astonishing, and we think he would have been run down at the first burst, only, from the vast extent of the gardens, when once he was fairly started, he could not well escape. He got in the Duke of Devonshire's kitchen-garden amongst the cabbages, and could not get out again. As it was, he kept dodging in and out, from tree to tree, running from one tent to another, in the hope of eluding his pursuers, but all in vain; they never left him for a minute, and, wherever he went, there were always some hundreds close upon his heels [...] Ultimately the poor, panting, Nepaulese Ambassador was caught, and carried off in a carriage, to be uncarted again at some future festive occasion for the amusement of Her Majesty's respectable subjects.[34]

Although intentionally amusing, the article actually bears a startling similarity to an account of the hippopotamus' own capture from *Household Words*:

> Nobody moved – not a green flag stirred; not a sprig trembled; but directly they entered, out burst a burly young hippopotamus-calf, and plunged head foremost down the river banks. He had all but escaped, when amidst the excitement and confusion of the picked men, one of them who had 'more character' than the rest, made a blow at the slippery prize with his boat-hook and literally brought him up by burying the hook in his fat black flank.[35]

Similarly, in reporting on his unusual eating habits, *Punch*'s suggestion of overcoming the prince's antipathy to western food with 'a few dozen of nice, fresh, live unopened oysters' directly insinuates that the Nepalese are both savage and animalistic, as, in eating live shellfish, 'they would be able to enjoy the luxury of killing and eating the natives.'[36] In contrast, the hippopotamus was elevated to the figure of muse by London society. A Hippopotamus Polka was composed soon after the creature's arrival and a play called *The Hippopotamus* opened in the Haymarket Theatre in August 1850. Declared by *The Times* to have 'little to recommend it beyond the popularity of its title,' the plot was centred on the theme of conjugal suspicion, 'by the circumstance that the jealous husband had been irritated by his wife's frequent visits to the wonder of the day.'[37] Although boasting a fight scene between the hippo and a lion as its finale, *The Times* concluded the play to be 'much too loosely connected [...] Absurdity is of course intended, but this need not be carried out in too fanatical a spirit.'[38]

34 *Punch* 19, 27 July 1850, p. 49. **35** *Household Words* 19, 3 Aug. 1850, p. 446. **36** *Punch* 19, 6 July 1850, p.23. **37** *The Times*, 13 Aug. 1850. **38** Ibid.

Both visitors were accorded audiences with Queen Victoria in the summer of 1850: Victoria attending the hippo at Regent's Park, and General Ranajee at a state banquet. The latter occasion allowed *The Times* considerable scope to speculate upon the state of Nepal, concluding that, although having been

> subdued, but not subjected to British dominion, and only partially to British influence [...] Since that period they have sate sullenly, and perhaps suspiciously, in their mountain home, debarred, like all other native Powers, from the pleasure of conquest by our supremacy.[39]

This assumption of superiority by *The Times*' journalist over a crown prince and his subject people is not only akin to the paternalistic condescension displayed toward the hippopotamus, but also reflects an attitude toward Imperial subjects that coalesced in the years following the Napoleonic Wars, 'one that was martial, nationalistic, paternalistic, moralistic, and racially pure.'[40] Indeed, less than a year prior to the opening of the Great Exhibition of 1851, Britain's showcase of how Imperial pomp and circumstance, modernized systems and innovatory technology could conquer the globe, a hippo and a Hindu exposed entrenched attitudes towards race and species that rivalled the more overt ideas of their American counterparts.

Although Britain may not have had an Agassiz or a Morton to consciously expound the polygenist doctrine, the humour of the popular press, which worked to objectify and gently ridicule the foreign 'other' (be it man or beast), nevertheless advanced the notion of kinship with a Nepali to be as implausible as that with a hippopotamus. Regardless of rank or bulk, both were seen to exist outside wholesale cultural assimilation and, as such, fulfilled the role of comic stooge. In June 1850, the entourage of the Nepalese prince was introduced to the new social craze of hot air ballooning, *The Times* reporting not only that, 'the height reached was understood to exceed anything yet accomplished in aerial navigation,' but that the show itself drew large crowds: 'the gardens were completely crowded with company, and for many years have not exhibited so gay a scene.'[41] In August 1850, *Punch* likewise mooted plans to present the

> Unparalleled Attraction [...] of Mr Green in his celebrated Fulham Balloon with the hippopotamus (of the Zoological Gardens) who has kindly lent his services for this occasion only. At the altitude of 200 feet above the level of Chelsea, Mr Green will descend from the car on to the back of the hippopotamus, and discharge a brilliant display of fireworks.[42]

39 *The Times*, 21 Jun. 1850. **40** Maya Jasanoff, *Edge of empire: conquest and collecting in the East, 1750–1850* (London, 2005), p. 308. **41** *The Times*, 24 Jun. 1850. **42** *Punch* 19, 3 Aug. 1850, p. 52.

Fig. 2: 'The next balloon ascent.' *Punch* 19, 12 Oct. 1850.

For 'seats on the back of the hippopotamus', readers were charged to apply at 'the Box Office of the Gardens.' *Punch* again proposed an aeronautical voyage for the hippopotamus in October, accompanying the article with a cartoon of a buoyant hippo.

On the same page as this article was the poem, 'The genuine prize song for Jenny Lind (at the service of Mr Barnum)'. Written in response to the two-year concert tour of America by the celebrated singer Johanna Maria Lind (1820–87), under the management of P.T. Barnum (1810–91), the poem is a wry reflection on the differences in attitude toward slavery between Europe and America (Britain finally abolished slavery in 1833, America would not do so until 1865). The poem ends with a 'comic' justification for the condoning of flogging and with the tagline that, in any case, coloured races are more animal than human:

No American can by his own fellow man
  Be disgraced with the stripes of the slave.
Man is sacred from blows – by the right of his nose,
  If it be not too broad and too flat;
Then you're licensed to thrash – then fall on with the lash –
  He's only a Nigger, and born to the cat!
La, la, la!
  Yes, a Nigger, and born to the cat![43]

Punch's specific mention of the American showman Phineas Barnum, founder of 'P.T. Barnum's Grand Traveling Museum, Menagerie, Caravan, and Circus' in

43 *Punch* 19, 12 Oct. 1850, p. 162.

1871 (which, after an 1881 merger with James Bailey and James L. Hutchinson, became 'P.T. Barnum's Greatest Show On Earth', and in 1888 the 'Barnum and Bailey Circus') further compounds this notion of divisions of race being akin to divisions of species. Barnum began his career as a showman in 1835 with the purchase and exhibition of a heavily disabled African-American slave woman, Joice Heth, who (Barnum claimed) had been the nurse of George Washington, and was over 160 years old.[44] As Barnum's troupe grew, he expanded his collection of human exhibits and, with the opening of 'Barnum's American Museum' in the 1840s, introduced the freak show to an all-too enthusiastic American audience. The exhibits themselves were often named after the animal their physiological appearance best corresponded to: Lionel the Lion-faced Man, the Alligator Boy, Hopp the Frog Boy, Koo Koo the Bird Girl, Priscilla the Monkey Girl, the Biped Armadillo, and the Leopard Family were just some of the 'bestial' exhibits on offer for public entertainment.[45] In thus animalizing the performers who sat on the other side of the platform to the audience, the latter were accordingly 'normalized', as Leslie Fiedler argues:

> A Victorian institution it is, like Victorian nonsense, intended to be finally therapeutic, cathartic, no matter what initial terror and insecurity it evokes. 'We are the Freaks,' the human oddities are supposed to reassure us, from their lofty perches. 'Not you, Not *you*!'[46]

Just as these freak shows served to advance polarizations between animal and human, black and white, normal and abnormal, they endorsed the disembodiment between spectator and spectacle, further compounding the 'normative' superiority of the (usually) white onlooker. Such clandestine polygenism was made palpable with Barnum's exhibition of a variety of African Americans with vitilgo, albinism, and microephaly. Proposing that these were the 'missing links' in the chain of evolution from monkey to white man, as Reiss argues, 'Barnum's later displays tended to reinforce racial boundaries as ineradicable: the "missing links" were presented as evidence for evolving racial science, which stressed the natural hierarchies of the races.'[47]

In much the same way, the nineteenth-century fashion for collecting exotica conferred ideologies of dominance. As Harriet Rivto contends, the longstanding practise of accumulating exotic animals for the purpose of display carried with it associations of 'triumphant individual enterprise as well of those of national prestige [...] that mirrored the spread of British commercial influence throughout the

44 Benjamin Reiss, *The showman and the slave: race, death and memory in Barnum's America* (Cambridge, MA, 2001), pp 1–2. 45 For a history of 'Freaks' and 'Freak Shows', see Robert Bogdan, *Freak show: presenting human oddities for amusement and profit* (London and Chicago, 1988); Leslie Fiedler, *Freaks: myths and images of the secret self* (London, 1978). 46 Fiedler, *Freaks*, p. 28. 47 Reiss, *Showman*, p. 42.

globe.'[48] The mid-Victorian craze for acquiring unusual flora, fauna, fossils and artefacts, and bringing them back to 'civilisation' was epitomized by the vast collections amassed by the burgeoning museums, such as the Victoria & Albert in London (established in 1852) and the Pitt Rivers in Oxford (founded 1884). Appropriating such items as the Elgin Marbles (taken from the Parthenon in Athens by Thomas Bruce, 7th earl of Elgin, and displayed in the British Museum in 1816) and exhibiting them for public edification figuratively classified such objects, and by association their country of origin, as subject to the long arm of Imperial authority. As such, museums not only fulfilled the role of educator but also of benign colonial benefactor: the action of displaying something – creating a spectacle out of it – having the effect both of exerting an attitude of paternalistic ownership over it, while simultaneously undermining the object's own code of identity. As Roslyn Poignant argues, 'the West's appropriation of colonial spaces in the course of the imperial enterprise was paralleled by the social construction and presentation of savage otherness in the showspace.'[49]

Parallels can undoubtedly be drawn between the creation of such museum collections and the long history of the transportation of indigenous peoples from their homeland to the 'civilized world' for the scrutiny of scientists and a curious public alike. The voyages of Bougainville and Cook in the 1760s and 1770s, for example, opened up the Pacific to Western eyes, expeditions returning with samples of their discoveries ranging from unusual plants and animals (including a parade of orang-utans, gorillas and chimpanzees) to Pigmies and Hottentots.[50] Stuffed specimens of aboriginal natives were subsequently exhibited in the glass cabinets of European collectors with the same sense of propriety as the display of exotic fauna with which they were surrounded.[51] This creation of spectacle under the auspices of scientific scrutiny was powerfully demonstrated in 1810 when Sartje Baartman, the 'Hottentot Venus', was brought to London from South Africa. Exhibited for commercial profit until her death in 1815, Baartman became an object of fascination due to her enlarged genitalia. She garnered enormous popular curiosity and coverage in the national press, underwent at least one anatomical examination during her lifetime (by the French anatomist Henri de Blainville, in 1815), and, upon her death, was subject to a post-mortem by the French anatomist Georges Cuvier whose account was published in the same year. Cuvier noted Baartman's 'simian' features and the 'brutish' appearance of her face, and conjectured that her large posterior was a feature shared by the apes.[52] In their

48 Harriet Rivto, *The animal estate: the English and other creatures in the Victorian age* (London, 1987), p. 206. **49** Roslyn Poignant, *Professional savages: captive lives and Western spectacle* (New Haven and London, 2004), p. 8. See also Thomas Richard, *The Imperial archive: knowledge and the fantasy of Empire* (London and New York, 1993). **50** Lynne Withey, *Voyages of discovery: Captain Cook and the exploration of the Pacific* (London, 1988). **51** Philip Blom, *To have and to hold: an intimate history of collectors and collecting* (London, 2002), pp 101–6. **52** Claude Rawson, *God, Gulliver and genocide: barbarism and the European imagination*

conclusions, both Cuvier and de Blainville echoed the arguments of Edward Long for a graduated scale of humanity, with Hottentots demonstrably at the bottom of the scale – more akin to the ape than the human species.

The concurrent arrival of Obaysach and General Ranajee in the summer of 1850, then, served to throw into relief an inherent attitude of social – and scientific – objectification toward both the native and the animal. The hippo, 'presented by H.H. the Viceroy of Egypt [and] exhibited daily from 1 to 5 o'clock, at the Garden in the Regent's Park,'[53] effectively embodied the commanding authority of the Empire. Subsequently given the moniker H.R.H (His Rolling Hulk) by *Household Words*, the spectacle created by, and the ensuing anthropomorphizing of the captive African creature further compounded the immutable authority of its white British observers. In much the same way, the attitude of benign paternalism and gentle condescension that accompanied the newspaper reports of Prince Bahadoor's visit functioned as a means by which the foreign native, regardless of his rank or wealth, might be manoeuvred into a position of relative inferiority. The display of both visitors compounded the objectification of each, and, subsequently, the bolstering of the host nation's self-regard.

Developments in evolution theory, however, scuppered these notional hierarchical boundaries between human and animal, race and species, effectively interconnecting all living organisms in a biologically ordained united network: the white Englishman could no longer so easily separate himself from the ape, let alone his fellow human. Charles Kingsley's popular Victorian fairytale *The Water Babies* appositely demonstrates both this growing acceptance of man's place within – rather than above – the animal kingdom, and also of the continued engagement of popular literature with contemporary scientific theories. The novel itself is self-consciously set in the age of evolutionary debate – Richard Owen, Thomas Henry Huxley, and Charles Darwin all being described as 'great men whom good boys are taught to respect,'[54] while the fairy godmother figure, Mother Carey, is the personification of natural selection, letting creatures 'make themselves.'[55] That Kingsley's work was regarded as a serious contribution to scientific debate is demonstrated by an appraisal in the *Anthropological Review*, which proclaimed that its publication 'marks the period of an epoch in our biological literature [...] [that] will open a new vista of contemplation wholly at variance with the habitual and unrefreshing thoughts which may have left feeble impressions on [...] plastic minds.'[56] Also included in this volume were articles by T.H. Huxley 'On man's place in nature', 'The relation of man to the inferior forms of animal life' by Charles S. Wake, 'Man and beast' by C. Carter Blake, and Charles Lyell on 'The geological evidence of the antiquity of man'.

(Oxford, 2001), pp 115–16, 127. **53** *The Times*, 12 June 1850. **54** Charles Kingsley, *The Water Babies*, ed. Brian Alderson (Oxford, 1995), p. 40. **55** Ibid., p. 149. **56** 'Kingsley's Water Babies', *The Anthropological Review* 1:3 (November 1863), 472–3.

Kingsley himself appositely embodied the perceived contradiction between developing scientific theories of the nineteenth century and traditional theocratic Creationism. A clergyman who openly accepted evolutionary theory, Kingsley was a practised marine zoologist and maintained correspondence at various times with a number of eminent scientists on both sides of the evolutionary debate, including Charles Darwin. Suffused with the imagery of Darwinist argument, *The Water Babies* worked to parody scientific and popular attacks on evolution theory as much as it lent its support to its promulgation. In particular, the novel ingeniously makes reference to the *hippocampus* debate that ignited between Richard Owen (1804–1892) and Thomas Henry Huxley (1825–1895). Owen was a longstanding supporter of the claim that, rather than being a primate, man instead formed a distinct subclass of mammals: this was revealed in particular, he argued, in the anatomy of the brain. To this end Owen maintained that the presence of a *hippocampus minor* was peculiar to humans and thus acted as a distinguishing trait that separated man from monkey. Huxley, however, an avid proponent of Darwin's evolution theory, asserted that apes not only possessed a *hippocampus minor*, but that it was at least as well developed as in man, if not better; thus proving, in other words, that *Homo sapiens* was a descendant of the higher apes. The argument was eventually won by Huxley who, at the Cambridge session of the British Association for the Advancement of Science in October 1862 dissected an ape brain to reveal its *hippocampus*, and thus Owen's error.[57] Kingsley not only repeated this incident in mock-serious parody in his novel but also, fascinatingly, employed the image of the hippopotamus as a vehicle for the adroit ridiculing of Owen's line of reasoning.

The hero of *The Water Babies* is Tom, a boy chimneysweep whose adventures in fairyland begin when he falls into a river and, waking up finds that he has turned into a be-gilled fish-boy. At one point in the story, Tom meets the bumptious Professor Ptthmllnsprts, naturalist and 'chief professor of Necrobioneopalaeonthydrochthonanthropopithekology.'[58] Holding 'very strange theories about a good many things'[59] he is described as having brought before the British Association the theory 'that apes have hippopotamus majors in their brains just as men have,' and the ridiculousness of basing the whole concept of evolution on such a minute biological point is starkly parodied:

> You may think that there are other more important differences between you and an ape, such as being able to speak, and make machines, and know right from wrong, and say your prayers and other little matters of that kind; but that is a child's fancy my dear. Nothing is to be depended

57 Huxley's theory of mankind's descent from an ape-like progenitor was conclusively stated in his book, *Evidences as to man's place in nature* (1863). 58 Kingsley, *Water Babies*, p. 81. 59 Ibid., p. 82.

> upon but the great hippopotamus test. If you have a hippopotamus major in your brain, you are no ape, though you had four hands, no feet, and were more apish than the apes of aperies. But if a hippopotamus major is discovered in one single ape's brain, nothing will save your great-great-great-great-great-great-great-great-great-great-great-greater-greatest grandmother from having been an ape too.[60]

Whether or not Kingsley was making a direct reference to Obaysach and the polygenist debates of a decade earlier, as well as to the Owen/Huxley tournament, is a moot point. The ingenious design of a Great Hippopotamus Test as a gauge for (un)humanness, however, accords with the comic potential of the animal that the popular press also drew upon to emphasize the 'otherness' of General Jung Bahadoor Ranajee. In this instance, however, Kingsley uses the image of the hippopotamus as a foil to ridicule those individuals for whom the designation of humanity also rested upon the designation of such biological minutiae as a *hippocampus minor* and a flat nose.

A final example of the utilization of the hippo as a comic foil for social purpose is the poem 'Lament of the Hippopotamus' from *Punch*. Published in 1851, it deftly exemplifies the extensive reconsideration of *Homo sapiens*' image of itself that evolution theory necessitated. Human grandiosity and self-glorifying eloquence is parodied by the articulate moping hippopotamus who is annoyed at being usurped in popularity by a baby Elephant. Played for laughs the poem certainly is; more fundamentally, it is also an animal talking to a human on an equal level, and even assuming a superior stance, mocking human fickleness and looking forward to the day when he, the beast, will once more reign supreme:

> Alas! for popularity – it is a fleeting flower,
> That buds and blossoms, and decays almost within an hour;
> 'Tis scarce a year ago that I was brought across the brine,
> To have a thousand worshippers bow daily at my shrine.
>
> To catch an early sight of me how fast the crowd would run,
> To see me sporting in my bath, or basking in the sun;
> Or in the gentle arms of sleep beneath a summer sky!
> No infant Hippopotamus was happier than I.
>
> They used to give me pails of milk – all genuine and true,
> But now amongst its tints I trace a skyish sort of blue:
> They used to come in carriages, and cabs and omni*bus*es,
> To see the Hippopotamus of Hippopota*mus*es.

60 Ibid., p. 83.

No longer do the Visitors to my abode repair;
In vain I take a sportive bath, or snuff the evening air:
To meet each other at my home they no longer appoint –
The hippopotamus is hipp'd – his nose is out of joint.

Oh where are all my worshippers – why am I left alone?
'Tis that a baby Elephant now occupies my throne.
The hippopotamus complaint, or mania, dies away;
Elephantiasis becomes the fever of the day.

But let the clumsy infant its triumph now enjoy;
The brute must quit its babyhood, and cease to be a toy.
Oh then farewell for ever, its glories and its charms!
It can't remain eternally an Elephant in arms.

'Tis then the Hippopotamus will reassume its sway,
Growing in popularity, as well as size each day.
Glories will light upon our race – the public will allot 'em us;
Then, hip! hip! hip! hip! hip! hurrah! hip! hip! for the Hippopotamus![61]

God-like no longer, then, nineteenth-century man stood revealed in all his beastliness. Far from upholding his place as the pinnacle of creation, it became increasingly perceptible as the century advanced that humans were simply animals by another name. Quantitative differences between species were reduced to qualitative differences of form as science turned from the heavens to the beasts for advances in understanding about *Homo sapiens*. The active engagement of the press and popular fiction with the development of evolution debates was both multilateral and incontrovertible, acting as the eloquent mouthpiece for the articulation of current opinion. This was particularly manifest in the developments in written humour. Although purposely light-hearted, the spectacle generated by the foreign visitors of 1850 effectively had the same consequence of racial and species 'othering' in the press as the more overt theories of polygenism being formulated in America at that time. As such, although Britain could boast neither an Agassiz nor a Morton, its hippo and Hindu were as good an indication of attitudes towards the granting of 'humanness' as measuring skulls or displaying 'freaks'. Latterly, the humorous hippopotamus was also used by the pro-evolutionist Charles Kingsley to disparage just those individuals for whom the designation of humanity rested solely on the supposed evidences of biological structure. Welcome or otherwise, the potential of the growling beast within found its natural imaginative home in popular literature, and writers did not fail to do it justice.

61 *Punch* 20, 7 June 1851, p. 226.

Edgar Allan Poe and the orang-utan

STEPHEN MATTERSON

> Of the orang-utan: it is a brute of a kind so singular, that man cannot behold it without contemplating himself.
>
> Georges-Louis Leclerc, Comte de Buffon (1707–1788)[1]

The title of this essay is an allusion to Poe's 1841 story, 'The Murders in the Rue Morgue', usually considered the first modern detective story. In it, Poe's Parisian detective, Monsieur C. Auguste Dupin, forerunner of Sherlock Holmes, is faced with what has since become a familiar problem in tales of detection, the closed-room mystery. In an apparently motiveless attack, a woman and her daughter have been brutally killed, and the body of the daughter thrust up the chimney of their fourth-floor apartment, which was locked from the inside. Dupin reasons that the murderer possessed tremendous physical strength and agility and combined these with a heartlessness bordering on animality: 'something altogether irreconcilable with our common notions of human action.'[2] The other major feature of the case is that witnesses overheard what they took to be the murderer, but were unable to agree on the language being spoken. With a bold imaginative move, Dupin combines these attributes and reveals the killer: an orang-utan that had escaped from the care of its owner, a sailor.

The figure of Dupin is of course crucial to the tale, and Dupin was to appear in two other celebrated Poe stories, 'The Mystery of Marie Rogêt' (1843) and 'The Purloined Letter' (1844). But it is interesting at this point to focus on the significance of Poe choosing the orang-utan as the killer. While the story is, as we shall see, central to Poe's representation of the scientist, 'The Murders in the Rue Morgue' is also Poe's intervention in an ongoing discussion at the time over what might be defined as 'human', a debate in which the orang-utan played a key role. Georges Cuvier, who is mentioned significantly in 'The Murders in the Rue

1 Quoted in Londa Schiebinger, *Nature's body: sexual politics in the making of modern science* (London, 1993), p. 75. 2 Edgar Allan Poe, *Poetry and tales* (New York, 1984), p. 422.

Morgue' had already played an important part in this, being one of the leading figures in the nineteenth-century delineation of racial difference. In 1816 he had dissected the body of Sarah Bartmann, the so-called 'Hottentot Venus' who had been abducted in 1810 and brought to Europe from South Africa to be exhibited as a wild beast in London. Public protests in England over this degradation led to her being liberated from the exhibition, and she was christened as Sarah Bartmann. She was however brought to Paris, where she was displayed in an animal show. There in 1815 she was examined by a team of zoologists and physiologists, Cuvier among them, who were eager to ascertain the nature of her being. As one member of the team, Henri de Blainville, put it, they wanted to compare her with 'the lowliest race of humans (the Negro) and the highest type of apes (the orang-utan).'[3] Bartmann died the following year, and Cuvier dissected her corpse. He subsequently wrote a now-notorious paper concluding that, judging by her ears, pelvis and heart, she was more like an ape than a human. Thus Bartmann was used to support Cuvier's earlier assertion that the negro race 'resembled apes'.[4]

While such science was enormously influential at the time (today we would say 'science'), and was used to support racialized slavery in the United States, what is especially relevant for reading Poe's story is Cuvier's analysis of how humans may be defined. The orang-utan functioned significantly in the debate that arose from this question, because its very existence apparently blurred the boundary between humans and apes. While the existence of the orang-utan had been known since the end of the seventeenth century, it became the focus of scientific debate in the nineteenth since it was then that the urgent questions that it raised became available to scientific scrutiny. During the seventeenth century, travellers to Malay had reported the orang-utan as human (its name means 'forest man' in Malay), and when it was first displayed in England in the early nineteenth century, it excited attention as a creature that seemed to be a combination of human and ape – in Poe's story, the similarity of the orang-utan's hand to a human hand fools the police. The existence of the orang-utan provoked a great deal of public interest (though it should be noted that until the mid-nineteenth century 'orang-utan' was not used specifically but was a generic term for 'ape'). In 1817 Thomas Love Peacock wrote a satire entitled *Melincourt* in which an orang-utan lives as a human, Sir Oran Haut-ton, who is elected to parliament. There was an outrageous attempt to mate a human female with an orang-utan in London, and even after 1844 when naturalists had definitively identified the orang-utan as a type of

3 Henri de Blainville, quoted in Schiebinger, *Nature's body*, p. 169. 4 Cuvier, quoted in Schiebinger, *Nature's body*, p. 258, n. 95. Sarah Bartmann's brain and skeleton were preserved, along with a wax model of her genitalia. The twenty-first century saw attempts to make amends for the degrading treatment she had received in the nineteenth: after a vigorous campaign her remains were returned to South Africa and were buried in a ceremony in August 2002.

ape, some elements of doubt remained in the popular imagination. This is evident from the exhibition of the so-called 'missing link' in London in 1846, sponsored by the American showman Phineas T. Barnum (it was in fact an actor in costume). The exhibition was advertised as follows:

> The long sought after link between man and the orang outang, which naturalists have for years decided does exist, but which has hitherto been undiscovered. It leaps, climbs and runs, with the agility of a monkey. It lays the cloth and sets a table with the sang-froid of a London waiter. Bows, lifts his hat and so on with the grace of a master of ceremonies; distinguishes colours, remembers what is said to it, goes through with military exercises and plays various games with an instinct and skill that would reflect honour on any gentleman.[5]

While there is no evidence that Poe was especially interested in the supposedly scientific theories of race and the use of these to defend slavery, as a Southerner and as someone with an interest in scientific debate, it is likely that he would haven been at least aware of such theories. But two things are apparent from 'The Murders in the Rue Morgue'. Firstly, Poe's main interest is in the intellectual process that leads to the discovery of the killer. Secondly, although the orang-utan may be seen to have blurred the absolute distinction between animal and human, it is clear that this blurring is only apparent. That is, the police are baffled because in their assumption that the perpetrator is human, they can only focus on the similarities between the ape and the human: the clues therefore point for them only to a human perpetrator, just as the witnesses who heard the killer assumed that human language was being spoken. Dupin's triumph of rational detection is brought about by his ability to see that the clues can indicate that the murders were committed by something non-human, and he therefore focuses on the differences rather than the similarities between how a human would behave and how an animal would.

As one of Poe's most widely-known stories, 'The Murders in the Rue Morgue' is most typically considered as one of his tales of 'ratiocination', distinct from his tales of terror. But this is a misleading distinction in some respects, because both the tales of ratiocination and the tales of mystery and imagination arise from Poe's strong interest in how we define and categorize the human, itself part of his lifelong concern with the nature of knowledge, with how we categorize, and what lies at the vulnerable and disruptible edges of our categories. As a snapshot of the status of science in the mid-nineteenth century, the example of the orang-utan is

5 P.T. Barnum, 'Advertisement for the Missing Link', London, 1846. Available online at: http://www.abc.net.au/rn/science/ss/stories/s1010792.htm, accessed 7 June 2006.

telling, not least for helping to understand the nature of Poe's work.[6] Poe was writing at a time when our modern distinctions between science, pseudo-science and entertainment were developing, and at a time when the individual branches of scientific thought were being defined. But the separation of science and spectacle were not absolute, as the case of Sarah Bartmann clearly illustrates. There was a large audience for scientific display and the exhibition of the extraordinary, and these were often most in evidence at the bazaar and the fairground. It is worth remembering that while mesmerism, named after Franz Anton Mesmer, was crucial to the development of psychoanalysis, it was in the early to mid-nineteenth century primarily a feature of the fairground. The huge interest in science and spectacle was both satisfied and excited in a masterly way by the aforementioned Phineas T. Barnum, Poe's almost-exact contemporary and the most famous American of the nineteenth century. In his American Museum in New York, founded in 1841, Barnum cleverly exploited the popular taste for wondrous science with a variety of exhibitions. There he held displays of scientific developments such as electricity, and exhibited a variety of creatures supposedly for their scientific interest. These included the duck-billed platypus, which generated widespread disbelief (though it was genuine), and the Fijian mermaid, which excited much interest (but was not). The American Museum's appeal to popular taste was both sensationalist and educational, such as when Barnum exhibited the conjoined twins from Siam, Chang and Eng. Barnum did employ a naturalist at the American Museum, Émile Guilladeau, whose job was supposedly to rule on the authenticity of the exotic exhibits – to his credit, he was not convinced by the Fiji mermaid. Barnum was also renowned for fooling the public with exhibits which appeared to be scientifically plausible. Indeed, some of these are especially interesting with regard to the dislocation of boundaries between humans and animals and between the races. In 1850, for instance, he organized an exhibit purporting to show a black man daily becoming more and more white, supposedly by ingesting an herb. While this was of course a hoax (a white man was being blacked up with progressively less make-up), it did potentially destabilize the racial boundaries which contemporary theorists had declared absolute.

While Barnum's primary objectives were entertainment and profit, the existence and popularity of his American Museum are both instructive for our reading of Poe. Like Barnum's exhibits, his stories are often combinations of pseudo-science, genuine scientific inquiry, and entertainment. In fact, Poe often

6 It is interesting that the orang-utan returns as the notional stage between humans and apes in American literary naturalism in the 1890s and early in the twentieth century, because of the acceptance of Darwinism as a social model in the work of the naturalists. The ship's master Dan Cullen in Jack London's 1908 story 'Make Westing' is compared to an orang-utan – intriguingly, the ship in this story is called the 'Mary Rogers', that is, the name of the murder victim in the real-life case on which Poe's 'The Mystery of Marie Rogêt' is based.

wrote hoaxes which, like Barnum's exhibits, exploited both the audience's gullibility and its capacity for wonder. The most notorious of these was 'The Balloon-Hoax' (1844), itself a variation on a Poe story from 1835, 'The Unparalleled Adventure of one Hans Pfaal.' Published in the newspaper the *New York Sun*, 'The Balloon-Hoax' was taken by readers to be the true record of a marvellous balloon journey which involved crossing the Atlantic in three days. Of course the journey was a complete fiction, but it is not hard to sympathize with the readers who believed it. Apart from being published through the medium of the newspaper, where a reader expects the factual, the feat did not seem unfeasible, especially to Poe himself, in an age of scientific wonders. Even today, readers are deeply divided over Poe's last major work, the long essay *Eureka* (1848). For some it has a serious scientific import which marks it as anticipating the work of both Freud and Einstein. For others it is the preposterous work of a fevered imagination. Perhaps our problem is in seeing it only as either of these things, whereas for Poe it may have been intended as both. That is, in 'The Balloon-Hoax', as in fiction such as *The Narrative of Arthur Gordon Pym* (1838), Poe's characteristic practice was to utilize a scientific mode of enquiry even when the ostensible subject of that enquiry was fictional or fantastic.

This practice is especially evident in one of Poe's most engaging essays, 'Maelzel's Chess Player', first published in 1836. For some time Poe had been fascinated by what had been widely publicized as an automaton that could play chess. This had been designed in the late eighteenth century by Wolfgang von Kempelen, a Hungarian-born inventor in the service of the Austrian empress Maria Theresa. Another inventor and showman, Johann Nepomuk Maelzel, bought the machine from von Kempelen and exhibited it throughout Europe and then in the United States in the 1830s. Poe wrote several features on von Kempelen, but it was this automaton which most fully engaged his interest. It is not hard to see why this should be so. Poe had long been fascinated by automata and he several times praised the ingenious and elaborate clockwork machines designed in the eighteenth century by the renowned Frenchman Jacques Vaucanson. The chess player seemed a reasonable development from such machines, and Poe was particularly fascinated because, like the orang-utan, it forced consideration of the attributes and limits of the human. The 1836 essay is an attempt to locate the principles on which the machine works. The paper is scientific in format and design, moving from observation of the machine to a consideration of human intelligence, to a list of seventeen points about the operation of the machine, and on to a conclusion where Poe deduces that this is not a machine but an elaborately designed box concealing a small man. The conclusion itself may not be very remarkable, but the actual argument of the essay demonstrates Poe's commitment to science not only as a subject but as a method of inquiry.

In spite of his conclusion, Poe takes very seriously the possibility of creating a chess-playing machine. With reference to the work of Charles Babbage, Poe points out that many tasks that we perceive to be characteristic of human intelligence are in fact mechanical because they are based on constants and on forms of repetition. For example, algebraic and arithmetical calculations are fixed and determinate, so that if the data is correctly inputted (as we would now say), a machine like Babbage's could be made that would do those calculations (Poe was certainly up to date with his science; Babbage had outlined the principle of the analytical engine less than two years before, in 1834). Poe points out that chess is a game defined by set rules taking place within a bounded space, and that the game's moves are unchangeable and essentially repetitive. Because of this, he suggests, it would be possible to develop a machine that would successfully play chess against humans (thereby anticipating *Deep Blue*'s defeat of reigning world champion Garry Kasparov by about 160 years). The automaton chess player was an exciting possibility to Poe precisely because it was possible. It took the clockwork inventions one stage further by applying the principle of repetition to cognition rather than to mechanics. Yet, by the same token, this is precisely how Maelzel's machine was revealed as a hoax and that it must contain a human player. Poe points out that if one were to design a machine that could play chess, it would do so consistently, responding predictably to particular moves, but the supposed machine played inconsistently. Poe demonstrates in the essay the crucial concept that repetition is the key to scientific understanding and successful experimentation. Chess as a formalized and defined mode of repetition means that a mechanical player is possible, but Maelzel's player cannot be a machine because the same results are not produced each time.[7]

It is obvious from a reading of 'Maelzel's Chess Player' that although science and entertainment were closely linked in Poe's time this should not lead us to assume that science was therefore trivialized. As noted above, the distinction we invoke today between science and pseudo-science was not so absolute in the mid-nineteenth century. Poe was typical among his contemporaries in taking phrenology very seriously, for instance, and we miss a great deal in his writing if we ignore that interest; that is, when Poe describes the heads of his characters he is indicating to us also, through the phrenological code, something more of their identity. The head of Roderick in 'The Fall of the House of Usher' (1839) is distinguished, Poe writes, by 'an inordinate expansion above the regions of the temple'.[8] In

7 One of Poe's recurrent interests was in cryptology. This is evident in 'Maelzel's Chess Player' insofar as he is considering the relation between cognition and communication, but it appears directly in a variety of essays, such as 'A few words on secret writing' (1841) and in several short stories, notably the well-known 'The Gold-Bug' (1843). Recent work on Poe has paid particular attention to his cryptology, and the techno-novelist Richard Powers indicated his literary indebtedness to Poe in the very title of his 1991 novel *The Gold Bug Variations*. **8** Poe, *Poetry and tales*, p. 321

phrenological terms this indicates his tendency towards abstraction. Poe is telling us, as one critic has put it, that Roderick 'seeks or perceives the truth beyond merely mundane phenomena.'[9] Mesmerism was even more important to Poe, since it opened the possibility for a sustainable mode of inquiry into the nature of the individual self. Phrenology might reveal the nature of character, but mesmerism interrogated the very essence of identity. Although both are now discredited, it is worth recalling that both were considered scientifically respectable in their day, and while mesmerism could be part of an evening's fairground entertainment, it was also a regular feature of Lyceum meetings and lectures.

Poe's interest in the boundaries of the self and in the relation between supposed opposites drew him naturally to mesmerism. The two sketches in which he considers it most directly as a subject were both published in 1845, 'The Facts in the Case of Monsieur Valdemar' and 'Some Words with a Mummy.' Poe's main concern here is with the boundary between life and death. Although this is ontologically and scientifically an absolute boundary, it may be a very fluid one in individual and imaginative terms. Indeed, in his fiction Poe frequently exploited his readers' anxieties over the perception that science was unreliable in distinguishing between life and death: a recurrent feature of his tales being premature burial. Yet Poe's interest here was not only in inducing terror in his readers. Frequently his characters exist within some boundary space between life and death, waking and sleep, reality and imagination: supposedly binary opposites which ought to have a clear demarcation but which no longer functions in these stories. Mesmerism is important for supposedly providing access to such boundary spaces, and when Poe invokes mesmerism directly this access leads to revelatory moments. Like Freud after him, Poe will seek to bypass consciousness in order to access some different form of truth. Mesmerism is important to Poe not because it necessarily led him to something new but that it seemed to him to confirm, and to do so scientifically, his belief that truth lies always beneath the surface of things and therefore requires an alternative mode of accession. His stories frequently occupy liminal spaces in which the reader is unable fully to determine the state of consciousness of the protagonist. The narrator of the 1838 tale 'Ligeia', for instance, tells the story of his first wife, essentially a life force that lives on by (literally) taking the life of another. The reader senses the narrator's unreliability given the uncertainty about his state of consciousness, to the extent that the story may seem like an uncensored dream. Similarly, critics have often pointed out that the opening paragraphs of 'The Fall of the House of Usher' seem to mimic falling asleep – or perhaps someone coming under the spell of the mesmerist – and the unforgettable story that emerges may come from the subconscious of the narrator. In effect, mesmerism is not so much

9 Beverly Voloshin, 'Transcendence downward: an essay on "Usher" and "Ligeia" ', *Modern Language Studies* 18:1 (1988), 21.

the subject of several of Poe's stories as it is the method by which those stories are told and how they reveal the truths of the self.

Poe dedicated *Eureka*, his 'essay on the material and spiritual universe', to the scientific traveller Alexander von Humboldt, but in the Preface he presents it 'to the dreamers and those who put faith in dreams as in the only realities.'[10] Poe's contemporaries might well have felt a contradiction between these two forms of dedication; one to a scientist, the other to dreamers. That we are less likely to do so today is mainly due to Freud, another believer in the truth of dreams who was also a scientist. Indeed, much has been written on the relation between Poe's work and the ideas that Freud developed half a century after Poe's death. The idea that Poe is Freud's precursor is a tempting one, though it can be overstated. In broad terms, they share attitudes and interests that many artists held in common; an interest in the uncanny and an anti-positivist stance. Freud himself, acknowledging the literary nature of many of his examples and analogies, genially pointed out that imaginative writers got there first: 'Not I, but the poets discovered the unconscious.'[11]

More specifically, Poe and Freud have in common a focus on the causes and effects of repression. Freud's famous insight in *Art and Literature*, 'When what has been repressed returns, it emerges from the repressing force itself',[12] could have come from a study of Poe's most well-known stories. Conversely, though, Freud makes possible a reading of Poe by providing us with a vocabulary and a cognitive map that were unavailable to Poe's contemporary readers. For some psychoanalytical critics the stories are read as Freud would read dreams, as revelatory narratives of the self. In this respect Poe's tales seem to form an easy access to Poe's own inner self, and many critics have taken their cue from D.H. Lawrence who declared that Poe 'is absolutely concerned with the disintegration-processes of his own psyche.'[13] The first major modern biography of Poe, by Joseph Wood Krutch, took a psychoanalytical approach, and Marie Bonaparte's two-volume psychoanalytic study was written with the approval of Freud himself.[14]

It is far more interesting, though, to look at the tales in the frame that Freud provides without reference to Poe's own psyche. For instance, 'The Fall of the House of Usher' readily lends itself to Freud's theories of repression. The twins Madeline and Roderick are the last of the Usher family and they live together in a gloomy isolated mansion, and there exists between them 'sympathies of a scarcely intelligible nature.'[15] If the family line – the House of Usher – is to be preserved, one of them must restore the disrupted identity boundary between them

10 Poe, *Poetry and tales*, p. 1259. **11** Quoted in Jeffrey Berman, *The talking cure: literary representations of psychoanalysis* (New York, 1987), p. 304, n. 40. **12** Sigmund Freud, *Complete psychological works of Sigmund Freud*, vol. 9, trans. James Strachey (London, 1959), p. 35. **13** D.H. Lawrence, *Studies in classic American literature* (1923; rpt. London, 1977), p. 70. **14** Joseph Wood Krutch, *Edgar Allan Poe: a study in genius* (New York, 1926); Marie Bonaparte, *Edgar Poe*, 2 vols (Paris, 1933). **15** Poe, *Poetry and tales*, p. 329.

and break from their intimacy that disallows sexual expression both within their relationship (because it is incestuous) and outside it (because it is thereby a betrayal of the living twin). It falls to Roderick to attempt to restore their identity boundary, and he may only move out of this familial sexual economy by effectively repressing Madeline. As the story develops he does this by declaring to the narrator that she has died. Since the narrator believes this to be true, he and Roderick place Madeline in a coffin which is then put into an underground vault that has a heavy iron door. In effect, Madeline has been repressed, and Poe's description of the vault suggests that she has been repressed by Roderick in an area of his subconscious:

> The body having been encoffined, we two alone bore it to its rest. The vault in which we placed it (and which had been so long unopened that our torches, half smothered in its oppressive atmosphere, gave us little opportunity for investigation) was small, damp, and entirely without means of admission for light; lying, at great depth, immediately beneath that portion of the building in which was my own sleeping apartment. It had been used, apparently, in remote feudal times, for the worst purposes of a donjon-keep, and, in later days, as a place of deposit for powder, or some other highly combustible substance, as a portion of its floor, and the whole interior of a long archway through which we reached it, were carefully sheathed with copper. The door, of massive iron, had been, also, similarly protected. Its immense weight caused an unusually sharp grating sound, as it moved upon its hinges.[16]

True to Freud's insight, the psychic energy Roderick expends in the act of repression goes to the entombed Madeline, which is used by her to awaken from her catatonic state and liberate herself, more than a week later, from both the coffin and the site of repression. Madeline's return is the return of the repressed. It destroys Roderick and it destroys her:

> As if in the superhuman energy of his utterance there had been found the potency of a spell – the huge antique panels to which the speaker pointed threw slowly back, upon the instant, their ponderous and ebony jaws. It was the work of the rushing gust – but then without those doors there *did* stand the lofty and enshrouded figure of the lady Madeline of Usher. There was blood upon her white robes, and the evidence of some bitter struggle upon every portion of her emaciated frame. For a moment she remained trembling and reeling to and fro upon the threshold – then, with

16 Ibid.

> a low, moaning cry, fell heavily inward upon the person of her brother, and in her violent and now final death-agonies, bore him to the floor a corpse, and a victim to the terrors he had anticipated.[17]

Such a Freudian reading of 'The Fall of the House of Usher' indicates the commonality of Freud and Poe, and illustrates how far Freud provides a frame for understanding and interpreting Poe. It also answers one of the questions that the story generates; that is, whether or not Roderick places Madeline in the coffin knowing that she is alive. For Freud and for Poe, the answer is that it does not matter. If Roderick is aware that she is living then his action arises from his conscious knowledge of his emotional need for freedom from her. If he is unaware of it, then the desire to suppress her has come from his unconscious. In both cases the psychic motivation is identical and the tale's psychological truth is intact.

Freud's observation on the return of the repressed can illuminate many of Poe's stories, which frequently involve forms of repression and the damage that repression can inflict on identity. In some of these stories where another person is repressed, Poe uses the literary device of the double. This is partially true of 'The Fall of the House of Usher', where the narrator notes the strong physical resemblance of the twins – in fact, a 1995 South African/Croatian television version of the story had the same actor playing both Roderick and Madeline.[18] The theme of the repressed double is most notably handled by Poe in 'William Wilson' (1837) in which, predictably enough, the narrator's killing of his double results in his own entrapment and death. However, while the psychological explorations of Poe's stories have most in common with Freud, his use of the double has a great deal of affinity with the work of Jacques Lacan. When Lacan represents desire as originating in lack that occurs because of the existence of the other, he is providing an insight into a recurrent theme in Poe. For both, the other means anxiety and loss, representing the separation of the self from the supposed sense of oneness, unity, with the world. Poe's stories often focus on this, but they also, as Lacan suggested in his 1957 seminar on 'The Purloined Letter', involve language in this dynamic; 'the unconscious is the discourse of the Other.'[19] Although Lacan's work on 'The Purloined Letter' is untypical, since he rarely addresses literary texts, it has been enormously influential in the development of critical theory, and theorists have repeatedly used Poe's stories to explore and to illuminate their concepts.

Poe as a precursor of Freud in exploring repression is evident in his fiction, but it is in his lecture-essay *Eureka* that Poe comes closest to anticipating another

17 Ibid., p. 335. **18** This series consisted of thirteen episodes and was a Dark Film Production made for Kushner-Locke and Allied entertainment. Each story was introduced by Christopher Lee. Available online at: http://www.mjsimpson.co.uk/leelostepisodeguide.html, accessed 7 June 2006. **19** John P. Muller and William J. Richardson (eds), *The purloined Poe: Lacan, Derrida and psychoanalytic reading* (Baltimore and London, 1988), p. 32. Lacan seminar trans. Jeffrey Mehlman.

aspect of Freud's ideas, notably his work on Eros and Thanatos. *Eureka* is the text in which Poe most clearly uses his understanding of science to explore and to present a theory of the relation between the material and the spiritual or, as we would be inclined to say after Einstein, between matter and energy. Poe delivered a first version of *Eureka* as a scientific lecture in February 1848, though when he enlarged and revised this for publication he insisted on it being called 'a prose poem', writing in the preface 'it is as a Poem only that I wish this work to be judged after I am dead.'[20] Perhaps this insistence was intended to warn the reader that the science in the essay was an imagined theory rather than a verifiable hypothesis, but as has often been pointed out, Poe's imaginative theories do anticipate both the 'big bang' theory of the universe's origin and Einstein's theory of relativity.[21]

In *Eureka* Poe states his belief that matter and energy are identical. As might be imagined, this for Poe can be nothing more than a belief, since both he and the age lack the scientific resources for any kind of verification. From this point Poe goes on to argue that the universe expanded from a single 'primordial' particle: 'The willing into being the primordial particle, has completed the act, or more properly the Conception, of Creation.'[22] Everything in the universe, all matter and all energy, partakes of the same identity and originates from the primordial particle; thus there is a unity in creation even though it has diffused from a single original point. However, the atoms of everything in the universe are attracted to one another and will, Poe suggests, eventually reform into one particle. That is, the universe expanded from a particle and is engaged in the process of returning to it. While Poe's belief is primarily to do with the cosmos, there is a personal aspect to it. He suggests that individuals as well as atoms seek unity and that individuality will collapse in the act of merging with others. This is where the link with Freud's Eros and Thanatos is most evident. Freud's Eros, the life-drive, corresponds with Poe's 'diffusion' while the death-drive, Thanatos, is postulated by Poe as the desire for unity that is also non-existence. In this regard, the desired unity of Roderick and Madeline Usher means a coalescence that is also death.

Poe believed strongly that in *Eureka* he had found the key to the universe, and although the science that he uses has been discredited, his overall theory has generally been verified. Yet it is also valid to see *Eureka* as Poe's most definitive statement about art as well as science. He argues that aesthetic pleasure derives from the satisfaction of our impulse to unity:

20 Poe, *Poetry and tales*, p. 1259. **21** See for example, George Nordstedt, 'Poe and Einstein', *Open Court*, 44 (1930), 173–80, and Juan Lartigue, 'Edgar Allan Poe and science: a cosmic poet' (2000), at http://www.poedecoder.com/essays/lartigue, accessed 7 September 2006. In his 'Commentary' on *Eureka*, Harold Beaver provides a useful overview of the contemporary science utilized by Poe. See *The science fiction of Edgar Allan Poe*, ed. Harold Beaver (London, 1976), pp 395–415. **22** Poe, *Poetry and tales*, p. 1277.

> The pleasure which we derive from any display of human ingenuity is in the ratio of the *approach* to this species of reciprocity. In the construction of *plot*, for example, in fictitious literature, we should aim at so arranging the incidents that we shall not be able to determine, of any one of them, whether it depends from any one other or upholds it. In this sense, of course, *perfection* of plot is really, or practically, unattainable – but only because it is a finite intelligence that constructs. The plots of God are perfect. The Universe is a plot of God.[23]

Reading *Eureka* today, it is most useful to see it not so much *as* science as being *about* science. Thus, we can see how Poe's reflections on scientific enquiry in his last major work have developed remarkably from those evident in the 'Sonnet – to Science' he published at the age of 20:

> Science! true daughter of Old Time thou art!
> Who alterest all things with thy peering eyes.
> Why preyest thou thus upon the poet's heart,
> Vulture, whose wings are dull realities?
> How should he love thee? or how deem thee wise,
> Who wouldst not leave him in his wandering
> To seek for treasure in the jewelled skies,
> Albeit he soared with an undaunted wing?
> Hast thou not dragged Diana from her car?
> And driven the Hamadryad from the wood
> To seek a shelter in some happier star?
> Hast thou not torn the Naiad from her flood,
> The Elfin from the green grass, and from me
> The summer dream beneath the tamarind tree?[24]

Here, on the face of it, Poe presents a fairly clichéd poetic theme, in which science is represented as the destructive enemy of the mythopoeic imagination. Science demythologizes our world and leaves it impoverished, leaves nature bereft of our enriching imaginings. Science is the enemy of the imagination. The poem, though, is slightly more complicated than it may at first appear. While science may be represented as a destructive force, it is also our contemporary in a way that the pagan gods of nature are not. In the poem's first line, with the phrase 'true daughter of Old Time' Poe is echoing a line from John Keats' ekphrastic 'Ode on a Grecian Urn' (1819), where Keats writes of the urn as a 'foster child of silence and slow time', a 'Sylvan historian' who 'canst thus express / A flowery tale

23 Ibid., p. 1342. **24** Ibid., p. 38.

more sweetly than our rhyme.'[25] That is, Poe sees science as the modern equivalent of the Grecian urn. While it expels pagan presences from nature and we feel the loss of these, it is nevertheless a vessel for truth and progress, a 'true daughter' of time. Science may leave the poem's speaker imaginatively bereft, but it represents progress and modernity that cannot be evaded. What Poe is in fact suggesting in the poem is that the relation between science and the imagination needs renegotiation. Enmity will lead only to the poet becoming increasingly archaic, turning to a past which while supposedly untouched by scientific thought is also more and more out of touch with contemporary reality.

Poe's reflections on science certainly became more informed and more sophisticated in the twenty years between his publishing this sonnet and his death. But he stayed true to the sentiment of the poem's first line in seeking continually for a mode of understanding that would combine the intuitions of the imagination with the rigour and rationality of science. His view of science, therefore, changes to bring it closer to the imagination and as his work developed Poe increasingly saw the scientific method as a form of intuition. That is, to be more than a record of reality, science needed to borrow from the imagination just as, to make it valid, the imagination needed to be grounded in the scientific. This intuition, or leap of imagination, would lead, as in the standard scientific method, to a hypothesis that is then tested repeatedly. But the imaginative leap was essential; it was this which would defamiliarize the world, it is this which would generate fresh hypotheses. Poe believed that in *Eureka*, the prose poem dedicated to a scientist, he had found the common ground between the scientific and the imaginative modes of apprehending reality. But in some important ways he had already done so, in creating his detective, Auguste Dupin.

'Poe invented the detective story in order that he might not go mad', claimed Joseph Wood Krutch in an influential judgment.[26] The detective is a figure of rationality who finds a reassuring order to what otherwise seems a menacingly arbitrary or chaotic reality. For Poe, dealing so often with the irrational and the chaotic, the detective story offered a stabilizing narrative grounded in the manageable real. It is often claimed that Dupin is essentially poetic as opposed to the scientific rationality of Sherlock Holmes. While it is true that Dupin relies more than Holmes does on the imagination, when his methods are examined it is clear that he is actually Poe's ideal of the scientist. In each of the three detective tales, his method is to draw a hypothesis from the material that he sees, and to do so without bringing distorting assumptions to the evidence. But the clues mean nothing without the liberating act of imagination which leads to the hypothesis. In 'The Murders in the Rue Morgue' Dupin makes the imaginative leap that lib-

25 John Keats, *The complete poems*, ed. John Barnard (London, 1979), p. 344. **26** Quoted in Philip Van Doren Stern (ed.), *The portable Poe* (New York, 1973), p. 330.

erates his thought from what everyone else has assumed; that the murderer was human. No-one else can solve the mystery of the two dead women because no-one else is able to make the leap of imagination that breaks them out of a set way of thinking. In effect, this closed-room mystery reflects on its own procedure, since all except Dupin are enclosed within a mode of thinking that could never lead to the killer. Poe's narration colludes with this, and locks us in the room, through the title of the story and through the first newspaper report of the killings, where they are called 'Extraordinary murders'.[27] By calling the events 'murders' the assumption of human agency has already been made and the characters of the story as well as its readers are locked into a system of thought within which the case could never be solved, no matter how rationally and scientifically one proceeds. As he starts his explanation of the mystery Dupin comments that the killings are 'altogether irreconcilable with our common notions of human action.'[28] In saying this he is demonstrating that he has already broken out of the closed-room thinking and that he will eventually solve the problem.

Once, reflecting on an exhibition of what we assume were still-life paintings, Poe made a comment conveying his sense of the function of the artist. 'In my view', he wrote, 'if an artist must paint decayed cheeses, [his] merit will lie in their looking as little like decayed cheeses as possible.'[29] It is not the job of the artist to mimic reality, to give us back only the surface of what we all see; the artist must reveal a deeper truth. The detective Dupin is a version of such an artist, who must see beneath the offered surface to find truth, who must demonstrate the capacity to re-imagine reality. For Poe that was the job of the artist. It was also the job of the scientist.

27 Poe, *Poetry and tales*, p. 405. **28** Ibid., p. 422. **29** Edgar Allan Poe, *Essays and reviews* (New York, 1984), p. 1330.

H.G. Wells and the imagination of disaster

DARRYL JONES

Empires at their height are haunted by images of their own inevitable downfall. Readers will certainly be familiar with the growing number of representations of the 9/11 tragedy in recent years, in a number of high-profile novels and, of late, films.[1] What may be forgotten in the wake of recent history is the ways in which the events of September 11 2001 were uncannily prefigured by a number of (even higher profile) Hollywood blockbusters of the 1990s, which featured the spectacular destruction of symbolic American architecture as central showcase effects: *Independence Day* (1994), in which the White House, the Empire State Building and the Statue of Liberty are destroyed; *Godzilla* (1998) shows the destruction of the Chrysler and Pan-Am/Met Life Buildings; *Armageddon* (1998), in which the Chrysler Building is destroyed; and *Deep Impact* (1998), in which the Chrysler Building is again reduced to ruins. These scenes provided a template and a hermeneutic for the inevitable cinematic character of the 9/11 attacks themselves, as Mike Davis suggests:

> the attacks of New York and Washington DC were organized as epic horror cinema with meticulous attention to *mise-en-scène*. The hijacked planes were aimed to impact precisely at the vulnerable border between fantasy and reality. [...] George W. Bush, who has a bigger studio, meanwhile responds to Osama Bin Laden as one *auteur* to another, with his own fiery wide-angle hyperboles.[2]

Musing provocatively – but very perceptively – on the fate of the World Trade Center as very much the greatest spectacle (the greatest *spectacular)* of the century so far, Jean Baudrillard writes:

1 Novels include Frederic Beigbeder's *Windows on the World* (2005), Jonathan Safran Foer's *Extremely Loud and Incredibly Close* (2005), Ian McEwan's *Saturday* (2005), Benjamin Kunkel's *Indecision* (2005), John Updike's *Terrorist* (2006). Films – of varying genres – include *Fahrenheit 9/11*, dir. Michael Moore (2004), *United 93*, dir. Paul Greengrass (2006), *World Trade Center*, dir. Oliver Stone (2006). 2 Mike Davis, *Dead Cities and other tales* (New York, 2002), p. 5.

> As for what should be built in their place, the problem is insoluble. Quite simply because one can imagine nothing equivalent that would be worth destroying – that would be worthy of being destroyed. The Twin Towers were worth destroying. One cannot say the same of many architectural works. Most things are not even worth destroying or sacrificing. Only works of prestige deserve this, for it is an honour.[3]

In spite of its own denials, America *is* a global imperial power, as writers as politically disparate as Régis Debray and Niall Ferguson have argued, and has been since at least the third decade of the twentieth century.[4] In what may reasonably be considered its late-imperial decadent phase, the United States of America has recently become even more than usually preoccupied with the End. Thus, the biggest-selling American authors of the twenty-first century are not, as might be expected, Stephen King or John Grisham, but Tim LaHaye and Jerry B. Jenkins, authors of the phenomenally successful *Left Behind* series of Evangelical Christian apocalyptic thrillers, with sales of 62,000,000 and counting.

Before the Americans, it was the British. The first phase of British imperial expansion, from the eighteenth century to the accession of Queen Victoria occasioned an enormous outpouring of these images, narratives and warnings of mass death and catastrophe. Plagues, fires, the destruction of cities, the fall of empires, last men, the end of the world: viewed in one light, these might seem to be the dominant cultural preoccupation of this 'long eighteenth century', their concerns predicating a great number of key and forgotten texts. Daniel Defoe must surely count as a major figure in this tradition: *Robinson Crusoe* (1719) ranks among the most significant 'last men' novels, its images of a sole surviving human being prefiguring a number of modern post-apocalyptic narratives, while *The Storm* (1703) and *A Journal of the Plague Year* (1722) depict the near-destruction of England by, respectively, tempest and disease. Toward the end of his career, Defoe capped these preoccupations with *The Political History of the Devil* (1726), in which the Prince of Darkness himself is revealed to have been an active agent across the course of human endeavour (as he remains for LaHaye and Jenkins, and presumably a large portion of their readership). Its first volume appearing rather prophetically in 1776, the year of the American Declaration of Independence, Edward Gibbon's magnificent *The Decline and Fall of the Roman Empire* (6 vols, 1776–88) is a work both historiographical but also monitory, outlining in fascinated detail the dissolution of the greatest of all empires as a pointed warning to his own contemporaries. If anything, the securing of the British imperium with the defeat of

3 Jean Baudrillard, 'Requiem for the Twin Towers', in *The spirit of terrorism and other essays*, trans. Chris Turner (London, 2003), p. 46. 4 [Régis Debray], *A modest proposal for a United States of the West, by Xavier de C***, prologue by Régis Debray*, trans. Joseph Rowe (Berkeley, 2004); Niall Ferguson, *Colossus: the rise and fall of the American empire* (London, 2005).

Napoleon in 1815 was accompanied by an even greater proliferation of such images. The liberal French philosopher C.F. Volney's *The ruins; or, Meditation on the revolutions of empires and the law of nature* (1791) became a major text for Romantic radicalism, its ideas animating perhaps the canonical English document of the Ruins of Empire, Percy Shelley's 'Ozymandias' (1817–18), as well as constituting an important part of the impeccable Enlightenment education which Mary Shelley gives to her monster in *Frankenstein* (1818). In 1826, Shelley made her definitive statement on the demise of humanity with her plague-novel *The Last Man* – just one of a number of works on this theme (and with this title) from around this time.[5]

But the height of British Imperial expansion was at the end of the nineteenth century, and it is this period and the decades to follow on which my essay focuses. The 1890s saw a number of novels in which the British Empire was introduced to its terrifying *Döppelganger*, novels of reverse imperialism in which They do it to Us: the titular protagonists of Bram Stoker's *Dracula* (1897), Richard Marsh's *The Beetle* (1897) and Guy Boothby's *Pharos the Egyptian* (1898) are all variously Oriental types (two undead Egyptians and an undead Wallachian) who move their operations to London, the Imperial Metropolis which they intend to overthrow. On a more secular level, the great criminal masterminds of the age – Professor Moriarty (Irish), Captain Nemo (Anglo-Indian), Fu Manchu (Chinese), Dr Nikola (Jewish, based in Port Said, but able effortlessly to pass for a Chinaman) – are all grudge-bearing colonial subjects plotting to undermine, if not destroy, the British Empire. Perhaps the greatest of all reverse-imperial novels is H.G. Wells' *The War of the Worlds* (1898), in which the Martians occupy the role the British themselves played in the same year, 1898, in the Battle of Omdurman – a colonial power committing mass slaughter by virtue of greatly superior technology (what the British Gatling gun was to the Sudanese assegai, so the Martians' heat ray was in turn to the Gatling gun). Wells, the greatest literary catastrophist of his or any age, also prophetically imaged forth, in his novel *The War in the Air* (1908), the destruction of Manhattan, and that in terms disturbingly familiar to all of us:

> The City of New York was [...] the largest, richest, in many respects the most splendid and in some the wickedest city the world had ever seen. She was the supreme type of the City of the Scientific and Commercial Age; she displayed its greatness, its power, its ruthless anarchic enterprise, and its social disorganization most strikingly and completely. She had long ousted London from her pride of place as the modern Babylon, she was

5 For studies of this, see Morton D. Paley, '*The last man*: apocalypse without millennium', in *The other Mary Shelley*, ed. Fisch, Mellor, and Schor (Oxford, 1993); Fiona J. Stafford, *The last of the race: the growth of a myth from Milton to Darwin* (Oxford, 1994), pp 197–231.

> the centre of the world's finance, the world's trade and the world's pleasure; and men likened her to the apocalyptic cities of the ancient prophets.
>
> For many generations New York had taken no heed of war, save as a thing that happened far away, that affected prices and supplied the newspapers with exciting headlines and pictures. The New Yorkers felt perhaps more certainly than the English had done that war in their own land was an impossible thing. In that they shared the delusion of all North America. They felt as secure as spectators at a bullfight; they risked their money perhaps on the result, but that was all.[6]

Nevertheless, come War does, with the destruction by aerial bombing of 'the centre of the world's finance, the world's trade'.

In 1941, George Orwell accused H.G. Wells of preaching 'the same gospel as he has been preaching for forty years, always with an air of angry surprise at the human beings who can fail to grasp anything so obvious.' He continued to argue that '[a]ll sensible men for decades past have been substantially in agreement with what Mr Wells says; but the sensible men have no power.'[7] Wells was certainly consistent in his thinking across a long and productive writing career from the 1890s to the 1940s, and most of the preoccupations of his earliest works – invasions, mass extinctions, eugenics and evolutionary speculations, all coupled with the perceived necessity of an overarching 'World State' governed by an ascetic elite of 'Samurai', an idea formulated at greatest length in *A Modern Utopia* (1905) – are reiterated and refined throughout the remainder of his writing. 'His thinking', John S. Partington notes in his detailed study of Wells' politics, 'developed in response to world events but did not change in nature at any time.'[8] In *The intellectuals and the masses* John Carey has written damningly of Wells' obsession with 'getting rid of people', as a characteristically modernist demophobic fantasy, but the issue is more complex and subtle than Carey's account allows.[9]

Towards the end of *The Time Machine* (1895), Wells' time traveller propels himself many years into the future to witness the end of the world:

> The horror of that great darkness came on me. [...] Then like a red-hot bow in the sky appeared the edge of the sun. [...] As I stood sick and con-

6 H.G. Wells, *The War in the Air*, ed. Patrick Parrinder (London, 2005), pp 128–9. **7** George Orwell, 'Wells, Hitler and the world state', in *The Penguin essays of George Orwell* (London, 1984), p. 195. **8** John S. Partington, *Building Cosmopolis: the political thought of H.G. Wells* (London, 2003), p. 2. **9** John Carey, *The intellectuals and the masses: pride and prejudice among the literary intelligentsia, 1880–1939* (London, 1992), pp 118–51. For another negative critique see Michael Coren, *The invisible man: the life and liberties of H.G. Wells* (London, 1993), though Coren's account is predicated rather damagingly on a fundamental hostility towards his subject marshalled in the name of vindicating G.K. Chesterton and Hillaire Belloc, the latter of whom in particular kept up an ongoing feud with Wells over faith and materialism throughout the 1920s and '30s.

> fused I saw again the moving thing upon the shore – there was no mistake now that it was a moving thing – against the red water of the sea. It was a round thing, the size of a football perhaps, or, it may be, bigger, and tentacles trailed down from it; it seemed black against the weltering blood-red water, and it was hopping fitfully about. Then I felt I was fainting, but a terrible dread of lying helpless in that remote and awful twilight sustained me while I clambered upon the saddle.[10]

The German physicist Rudolph Clausius' invocation of the principle of entropy in his own formulation of the Second Law of Thermodynamics (1850), with its implication that, the solar system being a closed system, the eventual heat-death of the sun and the end of all life on earth was a scientific inevitability, had a profound effect on later Victorian thinking, with images of a dying sun occurring in, for example, the works of Charles Dickens and Joseph Conrad, and here in *The Time Machine*. Images of 'Dead London', as we shall see, abound in late Victorian and Edwardian fiction, and it is worth remembering that what we have here at the very close of *The Time Machine* is an image of the end of *London*, for while the Time Traveller moves forward through vast gulfs of time, he stays completely still in space, never leaving his study – the whole novel, thus, never leaves London.

The same can be said for another great fin-de-siècle apocalyptic text, Conrad's *Heart of Darkness*, where Marlow tells his tale under a metaphorically dying sun:

> And at last, in its curved and imperceptible fall, the sun sank low, and from glowing white turned to a dull red without rays and without heat, as if about to go out suddenly, stricken to death by the touch of that gloom brooding over a crowd of men.[11]

Like Wells' Time Traveller, Marlow's tale of colonial atrocity in the Congo never leaves London, for the tale is told aboard a bark on the Thames, the centre of 'a waterway leading to the uttermost ends of the earth'.[12] As Marlow famously says of London: 'And this also […] has been one of the dark places of the earth.'[13] Dickens at the beginning of *Bleak House* also images forth an infinite London, stretching forward in time to the very ends of time, with its snowflakes 'gone into mourning, one might imagine, for the death of the sun' – but also stretching back to the dawn of time, 'as if the waters had but recently retired from the face of the earth, and it would not be wonderful to meet a Megalosaurus, forty feet long or so, waddling like an elephantine lizard up Holborn Hill.'[14]

10 H.G. Wells, *The Time Machine*, ed. Patrick Parrinder (London, 2005), p. 85. **11** Joseph Conrad, *Heart of Darkness* (London, 1993), p. 4. **12** Ibid., p. 4. **13** Ibid., p. 5. **14** Charles Dickens, *Bleak House* (London,

But images of a depopulated world were more than just a scientific abstraction for Wells and his contemporaries. Any reader of a newspaper from the mid-nineteenth century through to the 1940s and beyond would have been presented with innumerable accounts of mass deaths and holocausts. In 1877–88, the most powerful El Niño event in recorded history led to a great drought and a series of famines which swept across the globe from Egypt through India and the Far East, across the Pacific as far as Brazil. Further droughts followed in 1889–91, bringing famine to India, Korea, Brazil and Russia, and killing as much as a third of the populations of Ethiopia and the Sudan, and in a series of failed monsoons in the tropics and China in 1896–1902. With these famines came virulent plagues: malaria, bubonic plague, dysentery and cholera added millions to the deaths from starvation. Estimates for the total deaths from famine and its consequences from the 1870s to 1902 vary from 30 to 50 million people.[15] The beginning of the First World War in 1914 inaugurated what Eric Hobsbawm characterizes as 'the age of total war'. 'The lights are going out all over Europe,' British Foreign Secretary Edward Grey remarked on the night of the declaration of war between Britain and Germany. 'We shall not see them go on again.' At the same time, in Vienna, Karl Kraus responded to the declaration of war by embarking on a work called *The End of Humanity*. By the end of the war in 1918, the British dead numbered 800,000 (including half a million men under the age of thirty), the French 1.6 million, the Germans 1.8 million, the Americans 110,000. Total casualties are estimated around 10 million. This war also lead to the first truly modern genocide, the first systematic attempt to wipe out an entire population, the killing of 1.5 million Armenians by Turkey.[16] The casualties of the influenza pandemic of 1918–19 were more than double those of the War, approximately 20,000,000 deaths worldwide, including 850,000 in America.[17] The millions of deaths due to enforced starvation, social programming and forms of ethnic cleansing in the 1930s and the holocaust of the 1940s are too well documented to require further analysis here.

Two of Wells' major projects of the 1930s were for the cinema, working under the aegis of Alexander Korda to produce the screenplays for William Cameron Menzies' *Things to Come* and Lothar Mendes's *The Man Who Could Work Miracles* (both 1936). The latter was an extended reworking of a cheery 1899 tale of total world destruction:

> You see, when Mr Fotheringay had arrested the rotation of the solid globe, he had made no stipulation concerning the trifling movables upon its sur-

1991), p. 1. **15** Mike Davis, *Late Victorian holocausts: El Niño famines and the makings of the Third World* (London, 2001), pp 1–16. **16** Eric Hobsbawm, *The age of extremes: the short twentieth century, 1914–91* (London, 1995), pp 21–53. **17** For an account of the influenza pandemic, see John M. Barry, *The great influenza: the epic story of the deadliest plague in history* (London and New York, 2004).

> face. And the earth spins so fast that the surface at its equator is travelling at rather more than a thousand miles an hour, and in these latitudes at more than half that pace. So that the village, and Mr Maydig, and Mr Fotheringay, and everybody and everything had jerked violently forward at about nine miles per second – that is to say, much more violently than if they had been fired out of a cannon. And every human being, every living creature, every house, and every tree – all the world as we know it – had been so jerked and smashed and utterly destroyed.[18]

Things to Come is an inspired filming of the most detailed and ambitious of Wells' works of futurology, *The Shape of Things to Come* (1933).[19] *The Shape of Things to Come* continues Wells' long-standing interests in the specifics of megadeaths. In *The War of the Worlds* (1898), the 'heat ray' of the invading Martians brings wide-spread annihilation to the English home counties before they themselves are wiped out through contact with Earth germs:

> Had the Martians aimed only at destruction, they might on Monday have annihilated the entire population of London, as it spread itself slowly through the home counties. Not only along the road through Barnet, but also through Edgware and Waltham Abbey, and along the roads eastward to Southend and Shoeburyness, and south of the Thames to Deal and Broadstairs, poured the same frantic rout. If one could have hung that June morning in a balloon in the blazing blue above London, every northward and eastward road running out of the tangled maze of streets would have seemed stippled black with the streaming fugitives, each dot a human agony of terror and distress. [...] And this was no disciplined march; it was a stampede – a stampede gigantic and terrible – without order and without a goal, six million people, unarmed and unprovisioned, driving headlong. It was the beginning of the rout of civilisation, the massacre of mankind.
>
> Directly below him the balloonist would have seen the network of streets far and wide, houses, churches, squares, crescents, gardens – already derelict – spread out like a huge map, and in the southward *blotted*. Over Ealing, Richmond, Wimbledon, it would have seemed as if some monstrous pen had flung ink upon the chart.[20]

The War in the Air (1908) describes, as we have seen, the near-total destruction of New York from the saturation bombing of the German airfleet:

18 H.G. Wells, 'The Man Who Could Work Miracles', in *The complete short stories of H.G. Wells*, ed. John Hammond (London, 2000), p. 410. **19** For an excellent study of the making and significance of *Things to Come*, see Christopher Frayling, *Things to Come* (London, 1995). **20** Wells, *The War of the Worlds* (London, 1993), p. 97.

> The City Hall and Court-House, the Post Office and a mass of buildings on the west side of Broadway had been badly damaged, and the three former were a heap of blackened ruins. In the case of the first two the loss of life had not been considerable, but a great multitude of workers, including many girls and women, had been caught in the destruction of the Post Office, and a little army of volunteers with white badges entered behind the firemen, bringing out the often still living bodies, for the most part frightfully charred, and carrying them into the big Monson building.[21]

With the end of the war in the air comes a pandemic of 'The Purple Death', which precipitates the end of civilization:

> It is a universal dissolution. The fine order and welfare of the earth have crumpled like an exploded balloon. In five short years the world and the scope of human life have undergone a retrogressive change as great as that between the age of the Antonines and the Europe of the ninth century ...[22]

A new dark age is upon us, and the novel closes with a chilling evocation of a dead London:

> 'It's *London*,' he said.
>
> 'And it's all empty now and left alone. All day it's left alone. You don't find 'ardly a man, you won't find nothing but dogs and cats after the rats until you get round by Bromley and Beckenham, and there you find the Kentish men herding swine. (Nice rough lot they are too!) I tell you that so long as the sun is up it's still as the grave. I been about by day – orfen and orfen.' He paused.
>
> 'And all these 'ouses and streets and ways used to be full of people before the War in the Air and the Famine and the Purple Death. They used to be full of people, Teddy, and then came a time when they was full of corpses, when you couldn't go a mile that way before the stink of 'em drove you back. It was the Purple Death 'ad killed 'em every one.'[23]

Even the comic-satiric *Tono-Bungay* (1909) records a disastrous commercial venture to collect shiploads of quap, a substance so radioactive that it kills all who come into contact with it; while *The World Set Free* (1914) looks forward with chilling prescience to atomic warfare:

21 Wells, *The War in the Air*, ed. Patrick Parrinder, p. 139. **22** Ibid., p. 254. **23** Ibid., pp 267–8.

> For the whole world was flaring then into a monstrous phase of destruction. Power after power about the armed globe sought to anticipate attack by aggression. They went to war in a delirium of panic, in order to use their bombs first. China and Japan had assailed Russia and destroyed Moscow, the United States had attacked Japan, India was in anarchistic revolt with Delhi a pit of fire spouting death and flame; the redoubtable King of the Balkans was mobilizing. It must have seemed plain at last to every one in those days that the world was slipping headlong to anarchy. By the spring of 1959 from nearly two hundred centres, and every week added to their number, roared the unquenchable crimson conflagrations of the atomic bombs, the flimsy fabric of the world's credit had vanished, industry was completely disorganized and every city, every thickly populated area was starving or trembled on the verge of starvation. Most of the capital cities of the world were burning; millions of people had already perished, and over great areas government was at an end.[24]

The Shape of Things to Come has a World War raging from 1940 to 1950, a scenario which had also predicated Olaf Stapledon's history of the next two thousand years, *Last and First Men* (1930), which Wells had certainly read, and which opens with an exhortation to 'Observe now the history of your own epoch as it appears to the Last Men.'[25] If anything, Stapledon is even more addicted than Wells to megadeaths, poison gas, and germ warfare. Here, a series of devastating wars begins with the bombing of London by the French:

> In a couple of hours a third of London was in ruins, and half her population lay poisoned in the streets. One bomb, falling beside the British museum, turned the whole of Bloomsbury into a crater, wherein fragments of mummies, statues, and manuscripts were mingled with the contents of shops, and morsels of salesmen and the intelligentsia. Thus in a moment was destroyed a large population of England's most precious relics and most fertile brains.[26]

These wars end with the *total* destruction and depopulation of England, an event beyond even Wells' imaginings.

These visions of a Dead London, indeed, find reiteration across the fiction of a number of Wells' contemporaries in the field of popular fiction. In M.P. Shiel's *The Purple Cloud* (1901), Adam Jeffson, the Last Man in a world depopulated by a volcanic eruption of hydrocyanic acid, returns to his home city to find it littered with corpses:

24 Wells, *The World Set Free* (London and Glasgow, *c.*1921), pp 141–2. **25** Olaf Stapledon, '*Last and First Men' and 'Star Maker'* (New York, 1968), p. 17. **26** Ibid., p. 21.

> I came out into Farringdon Street, and at the Circus, where four streets meet, had under my range of vision four fields of bodies, bodies clad in a rag-shop of every faded colour, or half-clad, or not clad, actually in some cases overlying one another. As I had seen at Reading, but here with a more skeleton appearance: for I saw the swollen-looking shoulders, sharp hips, hollow abdomens, and stiff bony limbs of men dead from famine, the whole having the bizarre air of some *macabre* battlefield of marionettes fallen [...][27]

In Grant Allen's 'The Thames Valley Catastrophe' (1901) a volcanic 'fissure eruption' floods the Thames Valley with lava, which advances unstoppably towards London. The narrator, on a bicycling holiday, rushes back to London to rescue his family. London is destroyed, Manchester becomes the new seat of government, and the Thames Valley is transformed into 'The Glassy Rock Desert'. Cutcliffe Hyne's 'London's Danger' (1898) has a fire ravaging the city, killing half a million people and destroying the Bank of England and the Tower. The British economy is ruined, and the Empire crumbles. Manchester again becomes the seat of government. The socialist writer Fred M. White wrote a number of London catastrophes. In 'The Dust of Death' (1903), London suburbs are being jerry-built on top of the city's waste (like *Bleak House*, this is London built on disease, dirt and filth). A building project in Devonshire Park causes an outbreak of 'Label's diptheria' – over 7,000 are infected, and affluent London is evacuated. The polymathic Dr. Label insists that the waste can be sterilized by use of electricity, but his plan is rejected as too expensive: the plague changes the authorities' mind. In 'The Four Days' Night' (1903), London's petroleum storage tanks catch fire, and a cloud of impenetrably sooty smoke envelops the city, completely shutting out the light. The narrator, along with the aviator Sir Edgar Grimfen, takes to the skies in Grimfen's airship and sets off explosives in the atmosphere to dissipate the smoke. This causes rain, which does the trick. Fatalities amongst the poor are high – over 2000 children dead in the East End. The end of cities, and thus of industrial capitalism, is prophesied:

> Men fancied a city with six million corpses!
>
> The calamity would kill big cities altogether. No great mass of people would ever dare to congregate together again where manufacturers made a hideous atmosphere overhead. It would be a great check upon the race for gold.[28]

27 M.P. Shiel, *The Purple Cloud* (Lincoln, NE, and London, 2000), p. 111. **28** Fred M. White, 'The Four Days' Night', in *Science fiction by the rivals of H.G. Wells* (Secaucus, NJ, 1979), p. 449. The stories discussed here, by Hyne, Allen and White, are all collected in this extraordinary volume.

At the end of the story, a clean air act is passed. Finally, White's 'The Four White Days' is the most directly socialist of all. Here, an unparalleled blizzard paralyses London. Greedy capitalists buy up the stocks of coal, which they sell at grotesquely inflated prices. John Hampden, a radical MP, incites London's workers to march upon the coal stocks and seize them. The government, powerless to intervene in laissez-faire capitalism by taking control of coal supplies themselves, is powerless to stop the workers' march. The Trades Union movement wins the day.

These concerns stretch beyond London, often, as we have seen, bringing with them a perspective of a new global geopolitics. Edgar Wallace, easily the best-selling novelist of the period (in 1928 alone, with the exception of the Bible, one in four novels printed and sold in England was by Wallace),[29] turned inevitably to catastrophe fiction in his massive *oeuvre*. *The Green Rust* (1919) exploits post-World War I paranoia by having the defeated Germans turn to industrial chemistry: the villainous Dr. Van Heerden (concealing his German identity under a Dutch name), creates the eponymous substance with the aid of a group of disgraced scientists with the aim of causing global crop failure, precipitating both widespread starvation *and* economic collapse. Jack London's 'The Unparalleled Invasion' (1910) has the dubious honour of being perhaps the most comprehensive of all catastrophe narratives. London's story is set in the 1970s, by which time the population of China has increased dramatically to nearly one billion. The Western powers respond, unprovoked, to this perceived threat by bombing China with a combination of every known biological agent:

> The plague smote them all. [...] Had there been but one plague, China might have coped with it. But from a score of plagues, no creature was immune. The man who escaped smallpox went down before scarlet fever; the man who was immune to yellow fever was carried away by cholera; and if he were immune to that too, the Black Death, which was the bubonic plague, swept him away. For it was there bacteria, and germs, and microbes, and bacilli, cultured in the laboratories of the West, that had come down upon China in a rain of glass. [...] Such was the unparalleled invasion of China. For that billion of people there was no hope. [...] There was no eluding the microscopic projectiles that sought out the remotest hiding-places. The hundreds of millions of dead remained unburied, and the germs multiplied; and, toward the last, millions died daily of starvation. Besides, starvation weakened the victims and destroyed their natural defenses against the plague. Cannibalism, murder and madness reigned. And so China perished.[30]

29 David Glover, 'Introduction', in Edgar Wallace, *The Four Just Men*, ed. David Glover (Oxford, 1995), p. x. **30** Jack London, 'The Unparalleled Invasion', in I.F. Clarke (ed.), *The tale of the next great war, 1871–1914* (Liverpool, 1995), pp 267, 269.

In *Things to Come*, Wells' World War brings about the total breakdown of society, symbolically imaged forth by the collapse of Central London into the Thames in 1968: 'The bed of the Thames buckled up and the whole of the Strand, Fleet Street, Cornhill and, most regrettable of all, the beautiful St Paul's Cathedral of Sir Christopher Wren [...] collapsed in ruin and perished in flame.'[31] Wells recounts in detail the various attempts at biological and chemical warfare during this war, most particularly the use of 'Permanent Death Gas', discovered in 1934: 'as it evaporated it combined with oxygen to form a poison effective when diluted with fifty million times its volume of air.' Its use renders the area around Danzig uninhabitable for many decades:

> The murdered region was not re-entered, except by a few specially masked explorers, until after 1960, and then it was found to be littered with the remains not only of the human beings, cattle and dogs who had strayed into it, but with the skeletons and scraps of skin and feathers of millions of mice, rats, birds, and suchlike small creatures. In some places they lay nearly a metre deep.[32]

A by-product of Permanent Death Gas, the Sterilizing Inhalation, renders a generation of Japanese sterile when it is deployed in Chinese bombing raids on Tokyo and Osaka. More devastating still than the gases and the war is the pandemic which follows:

> Cholera and bubonic plague followed, and then, five years and more later, when the worst seemed to have passed, came the culminating attack by maculated fever. [...] Wind, water and the demented sick carried it everywhere. About half humanity was vulnerable, and so far as we know now all who were vulnerable took it, and all who took it died. [...] Maculated fever had put gas warfare in its place. It had halved the population of the world. [...] Where war slew its millions in a few great massacres, pestilence slew its hundreds of millions in a pitiless pursuit that went on by day and night for two terrific years.[33]

The trope of the 'Last Man', the sole surviving member of a destroyed human race, accompanies these pestilential and apocalyptic speculations: Wells' Time Traveller is one such Last Man, witness to the final destruction of life on earth, as is Mr Fotheringay the miracle worker, sole survivor in a world depopulated by his own actions. As Fiona Stafford, Marie Mulvey-Roberts and others have shown,

31 Wells, *The Shape of Things to Come*, ed. John Hammond (London, 1993), p. 226. **32** Ibid., pp 160–1. **33** Ibid., pp 207, 209, 213.

the Last Man is a characteristic figure of Enlightenment and Romantic discourse, the Last of the Race, the man who has outlived all his contemporaries. It is this interest in the possibilities for humanity of a greatly extended lifespan which is characteristic of what George Sebastian Rousseau has called 'The geriatric enlightenment'.[34] Rousseau enumerates literally hundreds of the eighteenth-century's prodigious long-livers, whose very fact was understood as an indicator of the progressive nature of humanity – as we progress towards enlightenment, our lifespan increases drastically. Death is, in principle, defeatable; we can live forever. Thus, at the very end of his *Enquiry concerning political justice*, William Godwin notes that, were his Enlightenment-anarchist system to be adopted, it would inevitably lead to 'Health and the prolongation of human life'.[35] Indeed, the first edition (1793) of *Political justice* contains the following passage, excised from subsequent editions:

> Let us here return to the sublime conjecture of [Benjamin] Franklin, that 'mind will one day become omnipotent over matter.' If over all other matter, why not over the matter of our own bodies? If over matter at ever so great a distance, why not over matter which, however ignorant we may be of the tie that connects it with the thinking principle, we always carry about with us, and which is in all cases the medium of communication between that principle and the external universe? In a word, why may not man be one day immortal.[36]

It was in part this idea of Godwin's which inspired Thomas Malthus to write his *Essay on the principles of population* (1798), subtitled 'On the speculations of Mr. Godwin, M. Condorcet, and others', though Malthus here addresses more directly Godwin's 'Of avarice and profusion', Essay II of *The enquirer: reflections on education, manners and literature* (1797). Malthus was writing avowedly within the context of the French Revolution, which, he believed, 'like a blazing comet seems destined either to inspire with fresh life and vigour, or to scorch up and destroy the shrinking inhabitants of the earth'.[37] 'Population,' he wrote, 'when unchecked, increases in a geometrical ratio. Subsistence increases only in an arithmetical ratio. A slight acquaintance with numbers will shew the immensity of the first power in comparison with the second.'[38] With population doubling and redoubling, perpetually draining the world's resources, Malthus believed, then a

34 G.S. Rousseau, 'Towards a geriatric enlightenment', in *1650–1850: ideas, aesthetics and inquiries in the early modern era*, vi, ed. Kevin L. Cope (New York, 2001), pp 3–43. **35** William Godwin, *Enquiry concerning political justice*, ed. Isaac Kramnick (London, 1985), pp 770–7. **36** Godwin, *An equiry concerning political justice and its influence on general virtue and happiness* (Dublin, 1793), ii, p. 393. **37** Thomas Robert Malthus, *An essay on the principles of population*, ed. Philip Appleman (New York, 2004), p. 17. **38** Malthus, p. 19.

greatly extended human lifespan would be devastating: people needed to die in large numbers to check the problem of runaway population. Thus began a small but significant paper-war between Godwin and Malthus. Godwin replied to Malthus in *Of population: an enquiry concerning the power of increase in the numbers of mankind, being an answer to Mr Malthus's essay on that subject* (1820): far from being overpopulated, Godwin argued that 'the earth is not peopled'; humanity is 'a little remnant widely scattered over a fruitful and prolific surface.'[39] It is this debate between Godwin and Malthus on the potential future of an overpopulated, or of a potentially depopulated world, that informs his daughter Mary Shelley's 1826 novel, *The Last Man*, the narrative of Lionel Verney, the sole surviving human being in a world destroyed by plague.

Malthusian population studies themselves contributed to the great nineteenth-century quasi-science of eugenics, the term coined by Sir Francis Galton to denote the attempt to increase the numbers of 'genetically superior' persons through forms of selective breeding, and given its most celebrated articulation in his *Hereditary genius* (1869). Like very many of his contemporaries, Wells was profoundly interested both in eugenics and in Malthusianism, which provided a subtler means of population control than the megadeaths to which he was also given. 'Malthus', he wrote in *Anticipations* (1901),

> is one of those cardinal figures in intellectual history who state definitively for all time, things apparent enough after their formulation, but never effectively conceded before. [...] Probably no more shattering book than the *Essay on Population* has ever been, or ever will be, written.[40]

The novel *When the Sleeper Wakes* (1899) has its hero Graham, a radical pamphleteer, waking from a coma of 203 years to find himself the ruler of an overpopulated world out of a Malthusian nightmare: London has 33 million inhabitants, one-third of whom are the slaves of the Labour Company. Wells' most powerfully eugenic work is *Men like Gods* (1923), which posits an earth 3000 years in advance of our own peopled by a small number of super-evolved human beings who have eradicated disease from their world partly through wiping out all disease-bearing animals. In 1937, Wells published the 'biological fantasia' *Star Begotten*, a sequel of sorts to *War of the Worlds* in which it is suggested that the Martians, who cannot launch a direct invasion as they have no resistance to Earthly germs, are instead firing cosmic rays at the Earth, which alter human genetics, slowly transforming us into a race of Martians. Uncharacteristically gifted individuals, it is suggested, are the products of this Martian genetic engineering.

39 Godwin, *Of population* (London, 1820), pp 15–16. **40** Wells, *Anticipations of the reaction of mechanical and scientific progress upon human life and thought* (Mineola, 1999), p. 162.

The most celebrated eugenic fantasy of the 1930s is of course Aldous Huxley's *Brave New World* (1932), in whose genetically-engineered dystopia Lenina Crowne, like any good citizen, constantly wears her stylish 'Malthusian belt', stuffed full of contraceptives (birth control was known in the 1930s as 'Neo-Malthusianism'), and which climaxes with the death by suicide of its very own Last Man, John Savage. Huxley's political treatise *Ends and Means* (1938) discusses the '"mechanomorphic" cosmology of modern science', in which 'The universe is regarded as a great machine pointlessly grinding its way towards ultimate stagnation and death', and records that 'Introduced suddenly to this mechanomorphic cosmology, many of the Polynesian races have refused to go on multiplying their species and are in the process of dying of a kind of psychological consumption.'[41] Also from the 1930s is *After Many a Summer* (1939), in which the clinician Dr Sigmund Obispo, researching longevity, becomes fascinated by the Fifth Earl of Gonister, an eighteenth-century alchemist who claims to have discovered the elixir of life by ingesting the raw entrails of carp. The novel closes with Dr Obispo tracking down the 200-year-old Earl to his ancestral home in England where he lives in a special cell in the basement. The side-effect of the elixir is reverse evolution: he has degenerated into an ape. In 1949, Huxley followed this with *Ape and Essence*, a rebarbative dystopia. Here, the biological agent glanders has largely depopulated America:

> The great Metropollis [*sic*] is a ghost town, [...] what was once the world's largest oasis is now its greatest agglomeration of ruins in a waste-land. Nothing moves in the streets. Dunes of sand have drifted across the concrete. The avenues of palms and pepper trees have left no trace.[42]

The remaining inhabitants of Los Angeles have degenerated into barbarism, worshipping Belial and burning books in order to bake bread.

These, I would argue, are characteristic Modernist concerns. Wells' generation had lived through profound alterations in thinking about time.[43] Scientifically, these included Charles Lyell's formulation of geological 'deep time' in the 1830s, which laid the foundations for modern geology and paleontology, and therefore for thinking about the archaic status of the Earth and its life; Einstein's theories of space-time (Einstein was on cordial terms with Wells); and the discovery by Edwin Hubble in 1929 of galactic red shift, which led to the big bang theory of the origin of the universe, and to speculations as to whether its expansion would

41 Aldous Huxley, *Ends and Means: an enquiry into the nature of ideals and into the methods employed for their realization* (London, 1938), pp 123–4. **42** Huxley, *Ape and Essence* (London, 1949), p. 45. **43** For an account of this subject which abuts onto my own – though its concerns are rather different – see Stephen Kern, *The culture of time and space, 1880–1918* (Cambridge, MA, and London, 2003).

increase indefinitely and thus lead, in accordance with the Second Law of Thermodynamics, to the inevitable heat death of the universe, black, frozen and remote; or, whether gravity would eventually overwhelm all other forces, making the universe contract back in on itself culminating in a satisfyingly apocalyptic 'big crunch'. Philosophically, the work of Max Nordau and Oswald Spengler on forms of degeneration and decline, and of Henri Bergson on time-flux, are also significant here. It is therefore understandable, perhaps, that in *Time and Western Man* (1927), Wyndham Lewis was to criticize the Western intelligentsia's misguided obsession with temporality. The same year saw the publication of J.W. Dunne's *An experiment with time*. Dunne, a pioneering aeronautical engineer, was inspired by the apparent premonitory quality of his own dreams, coupled with his reading of Wells and his understanding of the implications of relativity, to posit with great seriousness the notion of ' "*Absolute Time*", with an absolute past, present and future. The present moment of this "absolute time" must contain all the moments, "past", "present", and "future", of all the subordinate dimensions of Time.'[44] His own work, Dunne claimed, contained 'the first scientific argument for human immortality'.[45] After reading Dunne, Jorge Luis Borges wrote a series of essays on time across the 1930s and '40s: 'The Doctrine of Cycles', 'A History of Eternity', 'Time and J.W. Dunne', 'Circular Time', and finally 'A New Refutation of Time'. Reading Dunne also inspired J.B. Priestley to write his series of 'time plays' in the 1930s: *Dangerous Corner* (1932), *Time and the Conways* (1937), and *I Have Been Here Before* (1937).

The middlebrow novelist Warwick Deeping, author of the bestselling *Sorrell and Son* (1925), wrote a pair of time-travel novels, *The Man Who Went Back* (1940) and *Live Again* (1942) and also, most interestingly, the (long) short story 'The Madness of Professor Pye' (1935). Pye, an embittered scientist, discovers how to harness atomic power (which he calls 'On-Force') to create a lethal ray-gun with the intention of destroying the whole of Europe. He succeeds in laying waste to the English home counties before being blown up in his laboratory by the plucky Mrs Hector Hyde, 'both a gentlewoman', we are told, 'and a world-famous aviator'.[46] Pye's On-Force gun is one of a great number of death rays and doomsday weapons to figure in the period's fiction, from Wells' Martian heat ray to the 'One Ring' to rule them all, whose various adventures saw print from 1937 to 1955. The destroyed home counties here signal not only Deeping's obvious debt to Wells, but also his acknowledged admiration for the late-Victorian novelist Richard Jefferies, whose apocalyptic fantasy *After London* (1885) has London as a noxious swamp at the centre of a great lake that covers all of southern England as far as Oxford. This is really the stuff of Victorian imperial crisis – the English political

44 J.W. Dunne, *An experiment with time* (3rd ed. London, 1934), p. 157. **45** Ibid., p. vii. **46** Warwick Deeping, 'The Madness of Professor Pye', in *Two in a Train* (London, 1935), pp 30–82. My thanks to Mary Grover for providing me with this and other information about Deeping.

system collapses, and the country is invaded by an army of marauding Welshmen, bent on retribution for a history of colonial injustice. Outside Oxford, the Welsh army meets an Irish army on the march for similar motives. The two armies effectively cancel each other out, thus allowing the English a precarious survival. Deeping, a man of rebarbative eugenic views, shared this admiration for Jefferies with Q.D. Leavis – *After London*, it seems, offered Leavis the perfect literalized vision of her own anti-metropolitan animus: here, London is a poisonous swamp. The decomposing bodies of millions of dead Londoners have turned what was a great city into a toxic hell, 'deadly marshes over the site of the mightiest city of former days':

> The city of London was under his feet.
>
> He had penetrated into the midst of that dreadful place, of which he had heard many a tradition: how the earth was poison, the water poison, the air poison, the very light of heaven, falling through such an atmosphere, poison. There were said to be places where the earth was on fire and belched forth sulphurous fumes, supposed to be from the combustion of the enormous stores of strange and unknown chemicals collected by the wonderful people of those times. Upon the surface of the water was a greenish-yellow oil, to touch which was death to any creature. It was the very essence of corruption. [...] The earth on which he walked, the black earth, leaving phosphoric footmarks behind him, was composed of the mouldered bodies of millions of men who had passed away in the centuries during which the city existed.[47]

This is familiar imagery to us by now, though I should note that it is by no means confined to writers in popular genres: that great work of high modernism, T.S. Eliot's *The Waste Land* is in essence a poem of 'Dead London'. As Eliot writes in the fifth section of that poem:

> Who are those hooded hordes swarming
> Over endless plains, stumbling in cracked earth
> Ringed by the flat horizon only
> What is the city over the mountains
> Cracks and reforms and bursts in the violet air
> Falling towers
> Jerusalem Athens Alexandria
> Vienna London
> Unreal[48]

47 Richard Jefferies, *'After London' and 'Amaryllis at the Fair'* (London, 1939), pp 169–70. **48** T.S. Eliot,

Wyndham Lewis criticized James Joyce for his temporal preoccupations: 'I regard *Ulysses* as a *time-book*', he argued, in which 'the reader is conscious that he is beneath the intensive dictatorship of Space-time'.[49] Though it is dangerous to attempt to reduce the concerns of such a promiscuous novel to any one issue, it is worth noting here that *Ulysses* (1922) contains its moments of apocalypse, in which '*the End of the World*' is '*a twoheaded octopus in gillie's kilts, busby and tartan fillibegs*'[50] and Dublin is represented as a destroyed city at the end of an apocalyptic war:

> Dublin's burning! Dublin's burning! On fire, on fire!
>
> (*Brimstone fires spring up. Dense clouds roll past. Heavy Gatling guns boom. Pandemonium. Troops deploy. Gallop of hoofs. Artillery. Hoarse commands. Bells clang. Backers shout. Drunkards bawl. Whores screech. Foghorns hoot. Cries of valour. Shrieks of dying. Pikes clash on cuirasses. Thieves rob the slain. Birds of prey, winging from the sea, rising from marshlands, swooping from eyries, hover screaming, gannets, cormorants, vultures, goshawks, climbing woodcocks, peregrines, merlins, blackgrouse, sea eagles, gulls, albatrosses, barnacle geese. The midnight sun is darkened. The earth trembles. The dead of Dublin from Prospect and Mount Jerome in white sheepskin overcoats and black goatfell cloaks arise and appear to many. A chasm opens with a noisless yawn …*)[51]

As with so many of the examples we have seen of these late-Victorian and Modernist dead cities (in Dickens, in Wells, in Conrad), Joyce's apocalyptic Dublin carries with it distinctly colonial or imperial overtones, as there seems little doubt that this is in part a fictional rendering of recent events in Irish (and European) history.

Joyce's reading of the eighteenth-century theorist of cyclical time, Giovanni Battista Vico, was directly to inform the fictional practice of *Finnegans Wake* (1939), and that reading of Vico is the subject of Samuel Beckett's contribution to the volume of essays *Our exagmination round his factification for incamination of Work in Progress* (1929). Ernest Hemingway's first major novel, *The Sun Also Rises* (1926), takes its title from the most famous of all accounts of cyclical time, the book of *Ecclesiastes*:

> The sun also ariseth, and the sun goeth down, and hasteth to his place where he arose. […]
>
> The thing that hath been, is that which shall be; and that which is done is that which shall be done: and there is no new thing under the sun.

Collected poems, 1909–1962 (London, 1963), p. 77. **49** Wyndham Lewis, *Time and Western Man* (London, 1927), p. 100. **50** James Joyce, *Ulysses*, ed. Jeri Johnson (Oxford, 1993), p. 477. **51** Ibid., p. 555.

> Is there anything whereof it may be said, See, this is new? it hath been already of old time, which was before us.
>
> There is no remembrance of former things, neither shall there be any remembrance of things that are to come with those that shall come after.[52]

The doctrine of cyclical time, of course, also predicates the great aphoristic coda to high Modernism, the opening lines of 'Burnt Norton' (1935), the first of Eliot's *Four Quartets*:

> Time present and time past
> Are both perhaps present in time future
> And time future contained in time past.
> If all time is eternally present
> All time is unredeemable.[53]

In these preoccupations, which unite Eliot, Joyce and Borges with Wells, Priestley and Dunne, and with a host of popular novelists throughout the 1930s and beyond, we get a sense of a genuine 'popular Modernism' operating across traditional aesthetic and canonical divides.

'In my beginning', Eliot wrote in 'East Coker', 'is my end'.[54] In keeping with the doctrine of cyclical time I want to close, as Joyce did in *Finnegans Wake*, where I began. That is, with George Orwell. Orwell's *1984* begins 'It was a bright cold day in April, and the clocks were striking thirteen.' As well as signifying a mechanized view of time represented by the twenty-four hour clock, there is also the sense here of time being out of joint: thirteen o'clock is an *evil* hour. Furthermore, it is a time beyond the end of time, past the traditional end-time of twelve. Winston Smith, being tortured by O'Brien, is given a terrifyingly apocalyptic summation of his existence:

> 'If you are a man, Winston, you are the last man. Your kind is extinct; we are the inheritors. […] You are rotting away,' he said; 'you are falling to pieces. What are you? A bag of filth. Now turn round and look into that mirror again. Do you see that thing facing you? That is the last man. If you are human, that is humanity.'[55]

52 *Ecclesiastes*, 1:6, 9–11. **53** Eliot, p. 189. **54** Ibid., p. 196. **55** Orwell, *1984* (New York, 1950), pp 222, 224.

Surrealism, psychiatry and the science of the irrational

BENJAMIN KEATINGE

The Victorian era saw unparalleled material progress which was inspired chiefly by developments in science. The technology of the steam engine made railways possible, the potential of electricity was realized, many people enjoyed greater material comforts, great feats of engineering were accomplished and medical science advanced steadily. Anyone who belittles Victorian materialism must also reckon with the benefits brought to society by nineteenth-century technological innovations. However, the Victorian era is also associated with the social upheavals and injustices wrought by industrialization and mass-production. From being a largely agrarian economy, Britain and other European nations became urbanized with often shocking results in terms of poverty, pollution, squalor and disease. Technological progress came at the price. Workers were disenfranchized, a hypocritical moral code constrained the individual and the forces of reaction vied with political progressives in managing this traumatic period of transition.

The ambiguities of progress are reflected in Victorian attitudes to science and the place of scientific disciplines in the curriculum. A new recognition of the possibilities of scientific discovery mingled with a certain snobbish distaste for practical disciplines among the educational elites. The phenomenon of the 'two cultures', defined by C.P. Snow in his 1959 Rede lecture at the University of Cambridge, emerged during the Victorian era. While the gentleman of leisure might dabble in science as a gifted amateur, full professional recognition and educational status were slow to develop and the most prestigious educational route remained via public school to read classics at Oxford or Cambridge. So while Victorian scientific advances are noteworthy indeed (not least in terms of the impact of Charles Darwin's revolutionary *The Origin of Species* of 1859), the social position of science remained uncertain. And the technological advances which the Victorian age heralded were, for many, a mixed blessing.

With the outbreak of war in August 1914, the long nineteenth century came to an end. The triumphal optimism of the early months of war soon evaporated as German and Allied forces dug in for their first long winter in northern France. Victorian faith in material and moral progress, the assumption that ever increasing wealth and ameliorated social conditions would curb the worst excesses of industrialism and the hope that Christian optimism would continue to sustain the British imperialist project were all to vanish on the Western Front. The First World War was the first fully mechanized, industrial conflict where the full weight of technological know-how would be employed against the enemy. Material progress became, in the war economy, the appliance of technology for the purposes of mass slaughter. As the full scale of the war's death-toll dawned on its participants, it became clear that here was the nemesis of progress. Some contemporaries and many subsequent commentators were inclined to see, in this mass sacrifice of young men from many nations, a kind of collective insanity as if, within the regimen and discipline of army life, a wider derangement afflicted the combatants, a kind of delirious historical nightmare.

This view was taken by a group of iconoclastic young men who founded the Dada art movement in neutral Switzerland in 1916. Famously, the term Dada was chosen in an arbitrary fashion, according to one account by plunging a knife into a dictionary. The founder and lead protagonist of the movement, Tristan Tzara declared 'Dada means nothing'[1] and another member of the Zurich group, Hugo Ball commented 'What we call Dada is foolery, foolery extracted from the emptiness in which all the higher problems are wrapped, a gladiator's gesture, a game played with the shabby remnants […] a public execution of false morality'.[2] The movement's spontaneous eruption in the otherwise sedate city of Zurich in February 1916 amounted to an act of defiance against everything Western civilization stood for. At the Cabaret Voltaire, founded in a tavern in Zurich, the Dadaists organized a series of scandalous *soirées* at which they recited nonsensical poetry, sang noisily, declaimed radical manifestoes often causing riots and commotion in the process. The Dada movement relied on shock tactics for its effects; it tried to wake Europe up from the historical nightmare it was living through. On one level, Dada inclined towards nihilism, the rejection of all accepted codes and conventions, in art as in life. It was an anti-art movement which opposed the traditional reverence and high seriousness accorded to the Old Masters and much nineteenth-century art. However, it was also an affirmatory movement. Its very name, Dada, means 'yes yes' in Rumanian, the native tongue of the voluble Tzara. If, according to Tzara, traditional art 'is a pretension'[3] then it must be replaced by Dada which is not so much an art movement as a heightened level of consciousness which recognizes the redundancy of traditional values on a war-stricken continent.

Significantly, Dada was an international movement. Originating in Zurich, offshoots emerged in New York, Germany and Paris. Just as the Bolsheviks were

1 Hans Richter, *Dada: art and anti-art* (London, 1965), p. 35. 2 Ibid., p. 32. 3 Ibid., p. 35.

to initiate a movement towards world revolution in October 1917, so the Dadaists proclaimed the end of everything on an international scale. As the historian of Dada, Hans Richter comments:

> Dada not only had no programme, it was against all programmes. Dada's only programme was to have no programme [...] and, at that moment in history, it was just this which gave the movement its explosive power to unfold *in all directions*, free of aesthetic and social constraints. This absolute freedom from preconceptions was something quite new in the history of art [...] Unhampered by tradition, unburdened by gratitude (a debt seldom paid by one generation to another), Dada expounded its theses, anti-theses and a-theses.[4]

Nothing could be less scientific than this anarchistic, international revolt which had no goals, aims or ambitions except to show up the absurdity of a civilization intent on blowing itself to pieces. In a sense, Dada was the cultural expression of an historical malaise which the authorities in Europe failed to recognize or ameliorate.

The future leader of the Surrealist movement, André Breton had been conscripted into the French army in 1915 as a medical auxiliary after two years studying medicine in Paris. He treated shell-shocked soldiers at an army hospital in Nantes before being transferred to St Dizier, a neuropsychiatric clinic near Verdun. Here he worked under Raoul Leroy, a former colleague of Jean-Martin Charcot, a famous psychiatrist, whose treatment of hysteria had influenced the young Sigmund Freud. Breton immersed himself in the study of clinical psychology reading Freud in depth, as well as works by Pierre Janet and the pioneering German psychiatrist Emil Kraepelin. In his work as an army psychiatrist and through his own researches, Breton became familiar with a whole range of neurological and psychiatric abnormalities. Although he was to abandon medicine after the war, his training as a psychiatrist profoundly influenced his thinking. For Breton, the human mind contains depths and profundities which everyday existence tends to obscure. The extremities of war-induced psychic trauma revealed these inner workings in ways which made conspicuous the ambiguities of consciousness versus the unconscious, normal versus pathological thought processes. Although Breton was to devote his life to art and literature, his thinking about the human mind reflected modern psychological theory and was foundational to his aesthetic theories. Poetry, for Breton, would always be excavatory, a journey into the inner realm of the individual (and collective) psyche. As Malcolm Haslam claims '[t]o Breton and his friends it was not enough to write poetry about science or even inspired by science; the poet should be a scientist himself and his poetry

4 Ibid., p. 34.

a scientific investigation'.[5] Many of the techniques adopted by the Surrealists were to rely on a quasi-scientific method whether it was automatic writing, frottage, dream analysis or collaborative roleplaying. Breton's movement was relentlessly experimental and, while their techniques were designed to harness subjectivity over the more objective methods of science, the experimental method was one of the fundamental tenets of the movement.

Surrealism emerged from the confused debris of the Dada movement which had run out of steam in Paris by 1922. The word 'surrealist' was first used by the French poet Guillaume Apollinare in 1917 to describe his 'surrealist drama' *Les mamelles de Tirésias* produced in that year. The word caught on and was defined by Breton in his 1924 Surrealist Manifesto as 'pure psychic automatism by which is intended to express [...] the true function of thought', that is 'thought dictated in the absence of all control exerted by reason, and outside all aesthetic or moral preoccupations'.[6] Breton went on to argue that 'Surrealism is based on the belief in the superior reality of certain forms of association heretofore neglected, in the omnipotence of the dream, and in the disinterested play of thought'.[7] One of the earliest Surrealist texts, written by Breton, was entitled 'Introduction to the discourse on the paucity of reality' (1924) and Breton constantly stressed the need to look inward for an alternative and more authentic reality. With the horrors of war in the immediate past (and experienced directly by many Surrealist writers and artists), it is hardly surprising that Breton should emphasize the value of inner over outer reality.

Unlike Dada, which was absurdist in a self-consciously negative way, Surrealism aimed to surmount this negativity and posit new, surprising juxtapositions to heighten our sense of reality and possibility. In the popular mind, Surrealism is most closely associated with bizarre confluences such as that defined by the self-styled French poet Comte de Lautréamont as 'the unexpected meeting, on a dissection table, of a sewing machine and an umbrella'.[8] The iconic images of Surrealist artists such as Max Ernst, Yves Tanguy, Salvador Dalí and Joan Miró, all rely on the shock tactic of taking things from their customary context and giving them a new reality, a heightened reality which they termed 'surreality'. They deliberately adopted a kind of methodical madness which exploits our rather banal and habitual association of certain objects to certain situations and then transforms the objects and our visual experience of them by showing them in new and surreal combinations. As such, even the most outlandish surrealist image is a carefully calculated juxtaposition of elements which co-opts chance only secondarily to design.

We should not overlook, then, the element of calculation in the Surrealist enterprise. Breton and his followers started with the hypothesis that the uncon-

5 Malcolm Haslam, *The real world of the surrealists* (London, 1978), p. 27. **6** André Breton, 'First Surrealist Manifesto', in Patrick Waldberg, *Surrealism* (London, 1965), p. 11. **7** Ibid. **8** Ibid., p. 24.

scious has as great a role in human affairs as deliberate and purposeful conscious activity. They then devised a range of techniques and styles including painting, sculpture, different kinds of poetry, the novel and polemical manifestoes in order to test their initial intuition and to delve ever deeper into the possibilities of surreal experience. Of course, there is an absence of objectivity; they are committed to a given course of creative exploration and if the experiment fails, they will redesign their methods sooner than reformulate their hypotheses. However, it is clear from the diverse methods of the Surrealists that their enterprise was experimental in that some artistic/psychological techniques proved more successful than others in evoking the realm of surreality.

A good example of Surrealist participation in altered states of consciousness is 'The Possessions', a series of simulations of pathological mental states written jointly by Breton and Surrealist poet Paul Eluard and published in their 1930 book *The Immaculate Conception*. These texts aim:

> to prove that the mind of a normal person when *poetically* primed is capable of reproducing the main features of the most paradoxical and eccentric verbal expressions and that it is possible for such a mind to assume at will the characteristic ideas of delirium without suffering any lasting disturbance, or compromising in any way its own *faculty* for mental equilibrium.[9]

So the authors participate verbally in mental illness in a series of mediumistic psycho-styles which simulate successively Mental Debility, Acute Mania, General Paralysis, Interpretative Delirium and Dementia Praecox. As exercises in literary-psychiatric nosology (or the classification of diseases) they read convincingly as examples which could have been (but were not) lifted from a psychiatric textbook. And of course, Breton's first-hand experience, as a former psychiatrist, of these mental states adds to their cogency. In modern psychiatric terms these simulations deal respectively with dementia, manic depression, general paralysis of the insane, paranoid delirium and schizophrenia. A French psychiatrist, Alain Rauzy, who wrote a thesis on Surrealism and psychiatry, notes that, with the exception of 'Interpretative Delirium', all the 'Simulations' are convincing from a clinical point of view.[10]

By using the term 'The Possessions' to describe their verbal forays into mental illness, Breton and Eluard use the religious idea of 'being possessed' by the devil and imply a psychoanalytic idea, current at the time, of 'dispossession' (we would now use the terms derealization and depersonalization) to describe a loss of self and estrangement from reality. Breton's vocabulary implies that there is both a

9 André Breton and Paul Eluard, 'The Possessions', trans. Jon Graham in André Breton et al., *The automatic message*, trans. David Gascoyne, Anthony Melville and Jon Graham, *Anti-classics of surrealism* (London, 1997), p. 175. **10** As Anthony Melville observes in Breton et al., *The automatic message*, p. 152.

spiritual and scientific aspect to insanity and Breton's rhetoric, here and elsewhere, tends towards a valorization of madness, casting doubt on the therapeutic efforts of the medical establishment. Breton had great empathy with the insane regarding their immersion in delusional fantasies as enviable and amounting to privileged access to the inner realm he sought. As fellow Surrealist Salvador Dalí wrote, almost contemporaneously with 'The Possessions':

> I believe that the moment is near when, by a process of paranoiac character and activity of thought, it will be possible (simultaneously with automatism and other passive states) to systematize confusion and to contribute towards the total discrediting of the world of reality.[11]

By systematizing confusion, Breton and Eluard give new clarity to alternative mental states. Just as the Surrealist painter seeks to arouse our consciousness with bizarre juxtapositions, so a delirious language aims to wake the reader up and to reinvigorate the stale world of reality. Breton's anti-psychiatric bias was rooted in the conviction that the insane did not inhabit a lesser world than their saner compatriots; rather, they enjoyed the relative bliss of evading the paucity of post-war social reality which Surrealism sought to counter and enliven. In as far as Breton rejected conventional medical interventions for mental illness, he was anti-psychiatric; but his anti-psychiatry was scientifically informed and showed a remarkable empathetic grasp of the subjective experience of mental illness. As such, it is not entirely unscientific since it gives us an insight into the type of confusions that doctors and scientists try to remedy, but often fail to understand empathetically. Indeed, the nosological precision of 'The Possessions' owes much to Breton's medical training. If Breton was often iconoclastic in his valorization of madness and vocal in his condemnation of psychiatrists, it was at least an informed iconoclasm.

If, as Dalí suggests, Surrealism aimed at the 'total discrediting of the world of reality' on the psychological plane, on the political and social plane it aimed rather at the transformation of reality. Karl Marx had argued in his 'Theses on Feuerbach' (written in the 1840s) that 'The philosophers have only *interpreted* the world, in various ways; the point is to *change* it.'[12] Marx attempted to show that the internal contradictions of capitalism would inevitably lead to its collapse and replacement by a new egalitarian social order. The Surrealists, like Marx, believed in changing the world and most were committed to the communist cause. Indeed, during the early 1930s, the Surrealist journal, which Breton edited, was entitled *Surréalisme au service de la révolution* and Breton attempted to ensure that

11 Cited by Matthew Gale in *Dada and surrealism* (London, 1997), p. 291. 12 Karl Marx, 'Theses on Feuerbach' in *The German Ideology*, ed. C.J. Arthur, trans. W. Lough, C. Dutt, and C.P. Magill (London, 1970), p. 123.

all Surrealists maintained a commitment to the revolutionary cause, even trying to expel Dalí from the movement for alleged reactionary views. However, Surrealism itself was riven with internal contradictions and the tensions between the French Communist party and the Surrealists reflected these contradictions. Breton had argued in the First Surrealist Manifesto of 1924 for a reconciliation between dream and reality in which neither the realm of imagination nor the material world would be privileged. In a kind a clarion call he wrote 'I believe in the future resolution of these two states – outwardly so contradictory – which are dream and reality, into a sort of absolute reality, a *surreality*'.[13] Needless to say, many committed communists regarded such statements as bourgeois obfuscation and thus treated the Surrealists with suspicion. Ultimately, Breton was unable to reconcile his movement's focus on the inner world with Communism's interest in the material world so that, by the time of the Second Surrealist Manifesto of 1929, he was forced to acknowledge the difficulties of commitment to the cause. Making his apologies to 'certain narrow-minded revolutionaries', Breton pleaded that he did 'not see why we should abstain from raising problems of love, dream, madness, art and religion, provided that we consider them in the same light [as] Revolution'.[14]

Again we see the conflict between objective and subjective viewpoints. The Communists believed that, objectively speaking, history was marching towards proletarian revolution and that their role was to assist the process of inevitable change. Important Surrealist figures shared this view and one of Breton's closest supporters, Louis Aragon eventually split from Surrealism to commit himself fully to the Communist cause. However, the subjective and psychological tenets of Surrealism ultimately hindered its political strategies. Breton aimed at a total interpretation of reality which overcame subjective/objective categories so that Surrealism became a way of life and an all-encompassing mode or style. Transforming the world, for Breton, also meant transforming life on the subjective and individual level. Both Communism and Surrealism sought to raise people's consciousness, but in Surrealism it was a total, integrated psychic transformation which was required while for Communism it was the abolition of 'false consciousness' and an awareness of revolutionary possibilities that had the priority. Surrealism incorporated this latter aspiration into its broad agenda, but it denied it as a single and exclusive imperative. Breton's expansiveness is best summarized in the Second Surrealist Manifesto where he writes:

> From an intellectual point of view, it was and still is necessary to expose by every available means the factitious character of the old contradictions […] if only to give mankind some faint idea of its abilities and to challenge it

13 Breton, in Waldberg, *Surrealism*, p. 70. **14** Ibid., p. 78.

> to escape its universal shackles to some meaningful extent […] There is every reason to believe that there exists a certain point in the mind at which life and death, real and imaginary, past and present, communicable and incommunicable, high and low, cease to be perceived in terms of contradiction. Surrealist activity would be searched in vain for a motive other than the hope to determine this point.[15]

We see then, in Breton's rhetoric, a search for the reconciliation of opposites. His language is ecumenical, open to different modes of experience and understanding. He uses an image from scientific experiment in his 1932 *Les vases communicants* (translated as *Communicating Vessels*) to sketch the plural possibilities of Surrealism. The interlinked vessels of the science laboratory symbolize the intercommunication between imagination and reality, dream and fact. In spite of Breton's role as a high priest declaiming orthodoxy, Surrealism remained a broad church. Its inclusiveness was part of its creed holding open the possibility of reconciling different modes of human knowledge.

The First World War had demonstrated the destructive potential of modern weaponry. The war had seen the invention of the tank, the introduction of military aircraft, refinements in artillery, poisonous gas and the extensive use of the machine gun to deadly effect. While warfare has always evolved and incorporated new technologies and new tactics, it was clear that in the Great War, the power of man to kill, maim and destroy had reached a new level which cast in doubt the wisdom of progress which had characterized the nineteenth century. However, the omnipresence of the machine, in civilian life as in war, was something artists, writers and ordinary people were increasingly obliged to come to terms with. Major events in the nineteenth century, such as the Great Exhibition of 1851 in London, had demonstrated the transformative potential of new technologies and many Victorians took seriously the need to adapt and evolve accordingly. Movements such as Dada and Surrealism tended to embrace technology in their artistic productions and they accepted and worked with the vertigo of the modern and its mechanistic implications. In this they were preceded by the pre-war Italian Futurist movement which actively celebrated new technologies. Led by Italian poet Filippo Tomasso Marinetti and featuring important Italian artists such as Carlo Carrà and Giacomo Balla, this group was among the first European art movements to recognize that modernity involved a reorientation which rendered traditional artistic modes irrelevant. As its name implies, Futurism actively embraced the future. In their 1910 manifesto, widely distributed, they argued that 'a clean-sweep should be made of all stale and threadbare subject-matter in order to express the vortex of modern life – a life of steel, fever, pride and headlong

15 Ibid., p. 76.

speed'.[16] Accordingly, they adapted Cubist methods to paint modern machinery, such as the automobile or the aeroplane, and they captured mechanized bodies in motion in the 'headlong speed' of modern life.

Futurism preceded Dada, but it did not survive the war which dispersed or killed its main protagonists. While Dada tended to mock and deride the present, Futurism actively embraced it, making a virtue of necessity and urging a new wave of mechanistic art. Two major Surrealist artists – Marcel Duchamp and Francis Picabia – tapped into this turn to the mechanical. Duchamp's *Nude Descending the Stairs* (1912) is a set of kinetic planes which imply motion with geometrical and mathematical precision in the manner of the Futurist Balla. And Duchamp would maintain, throughout his artistic career, an obsession with the mechanical. We see this in his famous 'ready-mades' where he takes a functional everyday object, such as a bicycle wheel, a bottlerack or (in one notorious case) a urinal, and signs it as a completed work of art with minimal modification. His large-scale work *The Bride Stripped Bare by Her Bachelors, Even* (also known as *The Large Glass*) constitutes a mechanization of human desire with machine-like human figures sculpted in lead wire and mounted on glass. This mechanomorphic approach influenced Picabia who was in New York during the years 1915–16 with Duchamp. Struck by the dynamism and technological vibrancy of America, Picabia wrote:

> Almost immediately upon coming to America it flashed on me that the genius of the modern world is in machinery and that through machinery art ought to find a most vivid expression […] The machine has become more than a mere adjunct of life. It is really a part of human life […] perhaps the very soul […] I have enlisted the machinery of the modern world, and introduced it into my studio.[17]

In such paintings as *Parade Amoureuse* (1917) body parts are replaced by machine parts and desire itself is mechanized, an energy expressed through technology. There is a certain soullessness to these paintings which implies, perhaps, the dehumanizing tendencies of modern civilization. Their automated eroticism tends deliberately towards mechanical emptiness rather than emotional warmth. By contrast, paintings such as *New York* (1913) and *I See Again in Memory My Dear Udnie* (1914), painted after Picabia's first visit to New York in 1913, are more vibrant and share in the Futurists' celebratory enthusiasm for modern life.

Picabia is rivalled, amongst the Surrealists, only by Max Ernst in the protean variety of his artistic production. In a career spanning sixty years, Ernst adapted

16 Cited by Herbert Read, *A concise history of modern painting* (London, 1959) p. 110. See also Caroline Tisdall and Angelo Bozzolla, *Futurism* (London, 1977). **17** Francis Picabia cited by Caroline A. Jones, 'The sex of the machine: mechanomorphic art, new women, and Francis Picabia's neurasthenic cure' in Caroline A. Jones and Peter Galison (eds), *Picturing science producing art* (London and New York, 1998), p. 145.

and invented new techniques (such as collage, frottage, decalomania) to create a highly diverse *oeuvre* which, like that of Picabia, integrates the mechanical into the artwork. Originally from Cologne, Ernst saw action in the First World War in the German artillery (where he unknowingly shelled trenches occupied by poet Paul Eluard who was to become a close friend). Discharged in 1917, Ernst was initially more influenced by Dada than by the Expressionist movement of his native country. Arriving in Paris in 1922, Ernst entered wholeheartedly into the emerging Surrealist movement, painting (among other things) a famous group portrait of major Surrealists with the title *Le rendezvous des amis* (1922). Ernst alludes directly to the murderous potential of modern weaponry in mechanical paintings like *Hydrometric Demonstration* (1920) which has all the complex, sinister intricacy of Franz Kafka's demented machinery in 'In the Penal Colony'. Ernst's *The Massacre of Innocents* (1920) is another war painting which directly equates the vertigo of the modern (railway tracks, aerial flight) with its calamitous manifestation in the trenches. The famous *Celebes* (1921), which hangs in the Tate Gallery in London, provides another example of demented machinery, this time with a sexual subtext, which also shares in the metaphysical menace of de Chirico's early canvasses. All these paintings suggest that science and technology, far from enabling and emancipating mankind, have in fact been misused. The destructive consequences of this misuse, Ernst implies, must now be analyzed and incorporated by the artist. The psychological and the mechanical meet reflecting the psychic scars of war and the failures of modern civilization. What emerges in Ernst, and in other Surrealist artists, is a kind of sexualized violence which bears witness to psychic and historical trauma. The mechanization of sex and the technologies of the dreamscape reflect the psychological and historical burdens of the artist in the modern world.

Another Surrealist who explored the confluence of history and psychology is Salvador Dalí. His paintings inspired by the Spanish Civil War include *Soft Construction with Boiled Beans: Premonition of Civil War* (1936) and *Autumn Cannibalism* (1936), both paintings which combine disturbing images of sex and violence. Of all the Surrealist artists, Dalí is most closely associated with the ideas of Freud. Whether he is dealing with his own personal neuroses, or with wider historical issues, Dalí tends to exploit the unvarnished psychic realities which Freud claims underlie human nature. Thus in early masterpieces such as *The Great Masturbator* (1929) or *Illumined Pleasures* (1929), Dalí incorporates Oedipal themes as if transposed from the analytic session. Slightly later in his career, Dalí would develop his own personal artistic philosophy which he termed the 'paranoiac-critical method' by which he produced an iconography of 'irrational knowledge' and 'delirium of interpretation'.[18] Ignoring Breton's desire to see reality and imagination fused in a fruitful alliance, Dalí triumphantly foregrounds irrational

18 As outlined by Dawn Ades, *Dalí* (London, 1982), p. 119.

obsessions in blatantly sexual and Freudian imagery. Unlike other dream-like images produced by Joan Miró and André Masson which appear as spontaneous, automatic drawings, Dalí's method seems premeditated to the extent that he anticipates all the textbook complexes which Freud describes so vividly and uses them programmatically in his painting.

Of course, if Surrealism were to have a bible it would probably be Freud's *The Interpretation of Dreams* (1900). Freud himself acknowledged that he did not 'discover' the unconscious, but he did more than any other thinker to explore its mechanisms and implications for human behaviour. He is said to have been impressed by Dalí's famous *The Metamorphosis of Narcissus* (1937), but he had serious reservations about Surrealism as a movement. For Freud, who had trained as a medical doctor and whose initial researches were in the field of neurology, psychoanalysis was a scientific method which, if conducted properly, could cure the psychoneuroses, that is inhibitions and phobias caused by repression. Freud saw the psychoanalytic method as scientific and vigorously defended its therapeutic efficacy. One obvious dilemma for the psychoanalyst is the necessarily impalpable nature of the unconscious which, by definition, cannot be observed or measured and whose mechanisms must remain, in part, a matter of speculation. Freud claimed to demonstrate, via his own clinical work with hysterical patients, the dynamics of the unconscious which, he suggested, plays a far greater role in human affairs than had hitherto been thought. The Surrealists seized on these claims as an opportunity to discredit the 'real' and valorize dreams, imagination and reverie. But whereas Freud was interested in what he called the 'latent content' of dreams, their underlying meaning and relevance to the clinical condition of the patient, the Surrealists were just as interested in the 'manifest content', the form and mode of expression which the dream takes. Indeed, the chief innovation of the Surrealists was their co-option of this 'manifest content' and use of it in works of art and poetry. While Freud remained unhappy at what he saw as the opportunist appropriation and simplification of his ideas, artists like Dalí gleefully exploited this new found dream language. Dalí's work, in particular, is problematic in that it seems to fuse both the 'latent' and 'manifest' content of the dream by depicting the fragmented, illogical sequence of dreams along with pre-fabricated Freudian interpretations of them all on the one canvass. With Dalí, there is little left to be interpreted since the Freudian analysis has already been done by the artist.

In a sense, both Freud and Breton shared the same objective, the incorporation of the language of the unconscious into reality. Through psychoanalysis, Freud claimed, the patient could learn to harness his irrational side and emerge as a whole, integrated human being in what amounted to a psychological cure. Likewise, on the social and artistic plane, Breton also sought a new integration of previously bifurcated realms, the rational and the irrational. Dalí's obsessive pursuit of irrational knowledge was incompatible with this aspect of Breton's movement and this was one reason why Dalí pursued his own path as an artist after 1939. Freud, however, remained puzzled, if curious, about Surrealism. As he wrote in a letter to Breton in December 1932:

> Although I have received many testimonies of the interest that you and your friends show for my research, I am not able to clarify for myself what Surrealism is and what it wants. Perhaps I am not destined to understand it, I who am so distant from art.[19]

We see here the tetchiness of a man of science whose ideas have been modified and adapted in ways he could hardly have predicted. In W.H. Auden's words, Freud's thinking had become 'a whole climate of opinion' even before his death in 1939.[20] In retrospect, Freud's faith in the scientific validity of his analytic method appears misplaced, while the Surrealists' experimentation with Freud's ideas seems to be in keeping with the more speculative claims of psychoanalysis.

Members of the medical establishment in France were sceptical, at first, about Freud's ideas, claiming, on the one hand, that French psychologist Pierre Janet had substantially anticipated them and, on the other, maintaining the aloofness of empirical men of science against the new, speculative psychology. Breton and his associates were the first to advocate Freud seriously in France and the link with Janet was exploited in the interests of broadening this reception. The tensions and hostility between Breton's group and the medical establishment were reflected in the latter's scepticism regarding Freud and, of course, Breton's iconoclastic attacks on institutional psychiatry did not make his advocacy of the Austrian psychologist more palatable to the establishment. The reception of Freud in France constitutes a major chapter in the history of psychiatry and one which lies outside the scope of this essay. Suffice to say that within the French tradition, major figures such as Jean-Martin Charcot, Joseph Babinski and Pierre Janet had sought the psychogenesis of hysteria as a kind holy grail of psychiatry. Indeed, Charcot's famous demonstrations of hysteria at La Salpêtrière, in which he would present a female hysterical patient before a group of male medical students in a quasi-theatrical fashion, aroused controversy both within and without the medical establishment. Charcot's faith in the neurological basis of hysteria proved misplaced and both his French associates, Babinski and Janet, as well as Freud, grappled more successfully with the condition.

The concept of hysteria has a certain *fin de siècle* vagueness about it, something which scientists resented and attempted to rectify, but which the Surrealists

19 Sigmund Freud to André Breton, Appendix to André Breton, *Communicating Vessels*, trans. Mary Ann Caws and Geoffrey T. Harris (Lincoln and London, 1990), p. 152. **20** W.H. Auden, 'In Memory of Sigmund Freud' in *Selected poems*, ed. Edward Mendelson (London, 1979), p. 93.

found congenial. In an article by Breton and Louis Aragon, published in a 1928 issue of *La Révolution Surréaliste*, hysteria is championed as a potentially liberating condition just as 'The Possessions' had championed other types of mental illness. For Breton and Aragon:

> Hysteria is a more or less irreducible mental condition, marked by the subversion, quite apart from any delirium-system, of the relations established between the subject and the moral world under whose authority he believes himself, practically, to be. This mental condition is based on the need of reciprocal seduction, which explains the hastily accepted miracles of medical suggestion (or counter-suggestion). Hysteria is not a pathological phenomenon and may in all respects be considered as a supreme means of expression.[21]

Charcot's demonstrations (captured vividly in a well-known painting by André Brouillet in 1887) had subjected his patients to the curious gaze of medical onlookers in a certain victimization of the female subject. Breton and Aragon sought to restore a level of dignity to the unbalanced mind and saw creative potential in the anxieties of the hysteric. Freud's ideas lent a measure of authenticity and psychological cogency to neurotic distress which the medical establishment too readily saw in terms of symptoms and pathologies. The ailments of the neurotic, no less than those of the psychotic, found validation in Surrealist polemics.

In this respect, the trauma of the Great War was also a turning point in the annals of human psychology. While Freud had laboured in nineteenth century Vienna to demonstrate the role of the unconscious, it was not until the inter-war period that his ideas fully circulated, due, in no small measure, to the phenomenon of war neurosis. As Breton had seen at first hand, soldiers subject to the extreme stress of the battlefront frequently broke down making Freud's observations regarding repression and neurosis immediately observable in the traumatized soldier. The pioneering work of W.H. Rivers in Britain (brilliantly portrayed by Pat Barker in her *Regeneration* trilogy) demonstrated the relevance of Freudian thinking to military psychiatry and French doctors, such as Breton's mentor Raoul Leroy, were quick to see the psychodynamic model as a useful one. If the hardened soldier could exhibit the same symptoms as the Freudian hysteric, then new concepts and methods were needed to address the problem, rendered especially urgent by the exigencies of war.

In this broad context, Surrealism appears far more than just post-war disaffection, which Dada essentially was. The war, literally, changed everything. As

21 André Breton and Louis Aragon, 'The fiftieth anniversary of hysteria', trans. Samuel Beckett in André Breton, *What is surrealism?: selected writings*, ed. Franklin Rosemont (New York, 1978), p. 426.

Breton saw, it was necessary to change our modes of thinking and also, he maintained, change life itself by altering the basis of social cohesion. The psychological insights of Surrealism now appear more significant than their social insights, but in the inter-war period, both were intrinsic to the movement. Surrealism essentially dissipated as an active movement after the Second World War when the French cultural sphere was gripped by Sartrean existentialism. But Surrealism represented the spirit of the age in inter-war Paris, just as the existentialists seemed to capture the cultural mood after the Second World War.

A radical loss of innocence resulted from the two World Wars and Surrealism may be said to participate in a major reassessment of human motivations and potential. There is, nonetheless, a compelling naïvity to some of Breton's writings. In his 1928 novel *Nadja*, which describes certain arbitrary encounters in Paris between the author and a mentally unstable woman he names Nadja, Breton had insisted 'Beauty will be convulsive or will not be at all'.[22] In later writings, such as *L'amour fou* (*Mad Love*, 1937), he reiterated the essentially convulsive and irrational nature of creativity and of love itself. In their negotiations with psychiatry, the Surrealists show a marked disinclination to restore civilized and rational order to the human mind, especially given their belief in the positive, redemptive possibilities of irrational experience. Madness and love were, after all, their favourite themes. The Victorian faith in progress was based on the rational conviction that improved scientific methods and technology would better the lot of humankind. The Great War had shown this conviction to be false and the Surrealist movement, along with Dada, sought a new basis for saying 'yes, yes' to human civilization. As Breton saw, the old convictions were redundant. A new psychology and a new society were the only way forward. In this respect, the Surrealists were, perhaps, ahead of their scientific contemporaries, and while their approach to science may have been cavalier, it is clear that the Surrealists and the psychiatric community were in dialogue during the inter-war period. We see, then, how Surrealist science, in this superficially unscientific movement, permeated a whole climate of opinion in Western Europe and contributed to one of the most vibrant art movements of the twentieth century.

22 André Breton, *Nadja*, trans. Richard Howard (New York, 1960), p. 160.

Logic in literature: bilogic strands in the poems of Louis MacNeice

ROSS SKELTON

I have long been interested in the relationship between artists and scientists, but when I opted for a career in philosophy my concern about the connections between the two approaches to knowledge represented by science and art became dormant. After lecturing in logic and the foundations of mathematics for some years I decided to train as a psychoanalyst. My main difficulty in this new discipline was to learn how to suspend the highly structured thought patterns of logic in favour of the more fluid mental stance required in clinical work. This version of 'negative capability', a notion psychoanalysts have borrowed from the poet John Keats, requires that the analyst be able to tolerate uncertainty and anxieties without irritable reaching after reasons or facts. A much more 'literary' stance is needed for psychoanalysis.

In order to make headway with both logic and psychoanalysis, then, I divided them into two distinct worlds. This seemed to be successful, enabling me to pursue both disciplines separately, until one morning after I had just given a lecture on infinity and set theory and took up a copy of *The International Journal of Psychoanalysis*. Leafing through the reviews I was astonished to come across a book with the improbable title of *The Unconscious as Infinite Sets* by a Chilean author, Ignacio Matte Blanco. I spent the next few weeks absorbed in this extraordinary work and eventually met the author. After some initial disagreements, which were published in the *International Journal*, based on my being 'too logical' and him being 'too psychoanalytical',[1] we became friends and he was delighted when I presented him with some applications of his ideas in understanding the poems of Louis MacNeice. MacNeice's poetry and bilogic seemed to be made for each other and I would like to share this conviction with you in this essay.

Modern logic was pioneered by Bertrand Russell and Gottlob Frege in the early twentieth century, around the time when Freud's *Interpretation of Dreams*

1 See Ross Skelton, 'Understanding Matte Blanco', *International Journal of Psycho-Analysis* 65 (1984), 453–56, and Matte Blanco, 'Reply to Ross Skelton's paper', *International Journal of Psycho-analysis* 65 (1984), 457–60.

(1900) first appeared. This modern logic is sometimes presented as a depth grammar of the sciences and it is true that the whole of science and its associated mathematics can be spelt out using modern logic. On the other hand, psychoanalysis is thought by some to be a science and others are adamant that it is an art.[2] Today it remains uneasily poised between the neurosciences on the one hand and literature on the other, which is closer to patients' actual lived experience. This 'feeling-thought' of lived experience is addressed by bilogic, not only in studying schizophrenia but in more normal states. For Matte Blanco, our feelings are subjectively experienced by us as (potentially) infinite and, too dangerous for us to experience directly, they are (usually) filtered by finite networks of thought. Applying Georg Cantor's notion of mathematical infinity, together with some logical concepts, he constructs a bi-logic which captures the dual-track nature of feeling-thought. In this essay I will suggest that many of MacNeice's poems exhibit this 'bilogical' structure.

Central to Freud's theory of the unconscious is the notion that thoughts are merely shadows of our feelings and while 'feeling-thought' is hard to articulate, it is at the heart of the psychoanalytic process. It has been addressed by a number of authors, such as Melanie Klein and her disciple, Wilfred Bion (the 'analyst' of Samuel Beckett),[3] but more especially Matte Blanco who has made a special study of this subject through his notion of a double strand or bilogic operating in feeling-thought. The main idea of bilogic is that each mental experience has two intertwined strands: a classical logic of 'difference' (called 'asymmetric' logic), where contradictions are not tolerated, and a non-classical logic of 'similarity' (known as 'symmetric' logic) which tolerates contradiction, infinities and equivalence of part/whole. However, the key difference between the two logics is this: asymmetric logic recognizes individual people, whereas symmetric logic only recognizes similarities or classes of individuals. Broadly speaking, conscious processes are dominated by asymmetric (difference) logic and unconscious ones by symmetric (similarity) logic, but of course both logics are at work in any mental process at the same time.

Anyone familiar with MacNeice's poetry will know of his fondness for lists of different objects. In his revolt against the very abstract philosophy of Georg Wilhelm Friedrich Hegel, which he describes in his autobiography (*The Strings Are False*), MacNeice says that he was jubilant at escaping into difference: 'lots of lovely particulars; I suggest keeping generalizations out of it. Leave that to the chorus'.[4] The field of feeling-thinking requires us to simultaneously grasp similarities and differences, what MacNeice called 'the drunkenness of things being var-

2 See the entries on 'Science and psychoanalysis' and 'Existential psychoanalysis' in Ross Skelton (ed.), *Edinburgh International Encyclopedia of Psychoanalysis* (Edinburgh, 2006), pp 413, 152. **3** See Susan Isaac, 'The nature and function of phantasy', *International Journal of Psychoanalysis* 20 (1948), 148–60, and Wilfred Bion, *Second thoughts* (London, 1967). **4** Louis MacNeice, *The Strings are False* (London, 1965), p. 283. **5**

ious'.[5] In what follows, I argue that MacNeice and the psychoanalyst Matte Blanco share Hegel's vision of unity in difference as central to their thinking. MacNeice revelled in differences between things, especially words. In 'Letter from India', however, he describes 'words [that] had become snakes biting their tails': the asymmetry of language, in a sense, had been 'subverted'.[6] MacNeice's line contains an allusion to the famous dream of the nineteenth-century chemist Friedrich August Kekule, who dreamt of this symmetric image while researching the molecular structure of benzene.[7] He interpreted his dream of a snake with its tail in its mouth to mean that the structure was a closed carbon ring.[8] This context gives MacNeice's phrase from the same poem ('An India sleeps beneath our West') an added poignancy for it hints at a collective mind underneath our Western civilization. MacNeice's poem 'The Death-Wish' has the lines: '[...] all our thinking gurgles down / Into the deep sea which never thinks',[9] which to me suggests a medium without differences, whereas the detail of the shore involves difference, the asymmetry of the individual mind. The sea for MacNeice may then be said to represent symmetric being, while the shore is asymmetric.

II

Psychoanalysis has long been familiar with thought processes which are overrun with feeling, and the phenomenon is often referred to by the term 'all or nothing' thinking where thought is almost drowned in emotion.[10] For example, the analyst mentions one point of similarity between a patient and her mother causing an explosion of anger. The patient experiences sharing even one quality with mother as being confused with her and thus to be the same as her. Mother, over the years, has become 'all bad' and so to have a single point of similarity with such a person is to become 'all bad' too. From the observer's objective standpoint we could say that such experiences were merely 'finite but total' but from the subjective point of view they are infinite. Someone with a bad anxiety attack is unlikely to say that they had a 'finite but total' feeling of terror and we would be suspicious of someone who claimed that they were 'in love' in a finite way. This suggests that our subjective feelings have a kinship with infinity. Experiences of love, hate, tenderness and beauty in the world or even the more prosaic joys in everyday life have a 'total' or infinite quality.

Matte Blanco brings Cantor's notion of mathematical infinity to bear on the nature of feelings. Most people already have the familiar notion of infinity, that it involves the idea of infinite duration or quantity, especially in space or time.

Louis MacNeice, 'Snow', in Peter McDonald (ed.), *Collected poems* (London, 2007), p. 24. **6** Ibid., p. 295. **7** See Carl G. Jung, *Man and his symbols* (London, 1964), p. 38. **8** Ibid. **9** *Collected poems*, p. 199. **10** For a more detailed discussion of this point see Ross Skelton, 'Generalization from Freud to Matte-Blanco', *International Review of Psycho-analysis* 17 (1990), 471–74.

Children are very interested in the spatial idea of the universe 'going on for ever', that is before they are domesticated into the adult's conscious ways of the finite. In his poem 'Nature Notes' the sea for MacNeice is described as being 'capable / Any time at all of proclaiming eternity'.[11] In his memoir he recalled a time at the age of two or three when he saw yachts sailing on Belfast Lough:

> [F]or some odd reason, a, perhaps, Jungian archetypal reason, I assumed that these sailing boats were moving in a westward direction, which meant going towards the head of the Lough, towards Belfast itself, would be going on forever in that direction like Columbus. I didn't realize, being only about two or three, that the Lough came to an end and they couldn't sail on forever.[12]

The notion of a space or a world without end fits in well with symmetric being for, as its relations are reversible, there can be no ordinary notion of space. A similar point holds in relation to time: the reversibility of relations in symmetric being disallows any serial ordering of time and thus timelessness is a chief feature of symmetric being. This reminds us of MacNeice's 'timeless' moments, as expressed in the phrase 'Time was away and she was here' in 'Meeting Point', to mention just one example.[13] Symmetric being – the One, indivisible, unified spiritual reality put forward by the pre-Socratic philosopher Parmenides who believed we cannot know this world – is indivisible, timeless and spaceless. It is my belief that the poetry of MacNeice also has an infinite, spaceless and timeless, core.

Cantor, who discovered the mathematical infinite(s), called them 'Alephs' after the first letter of the Hebrew alphabet. Instead of entering into the complexities of Cantorian infinites, however, it might be helpful to quote from Jorge Luis Borges' short story 'The Aleph'. The notion of infinity is highly abstract and Borges manages to capture the spirit of the idea visually here:

> I arrive now at the ineffable core of my story. And here begins my despair as a writer. All language is a set of symbols whose use among its speakers assumes a shared past. How, then, can I translate into words the limitless Aleph, which my floundering mind can scarcely encompass? Mystics, faced with the same problem, fall back on symbols. [...] The Aleph's diameter was probably little more than an inch, but all space was there, actual and undiminished. Each thing (a mirror's face, let us say) was infinite things, since I distinctly saw it from every angle of the universe. I saw the teeming sea; I saw daybreak and nightfall; I saw the multitudes of America ...[14]

11 *Collected poems*, p. 549. 12 Louis MacNeice, 'Autobiographical talk', in Alan Heuser (ed.), *Selected prose of Louis MacNiece* (Oxford, 1987), p. 267. 13 *Collected poems*, p. 184. 14 Jorge Luis Borges, *Collected fictions*, trans. Andrew Hurley (New York, 1998), p. 27.

This example captures the spirit of Cantor's infinite, but now let us turn to the letter of his theory. Cantor's infinity is a kind of untamed branch of modern mathematics which has generated much controversy around the fact that Cantor supposes infinite processes can be completed. This is because it has also generated a number of virtually unsolvable mathematical paradoxes, including Russell's, according to which the set of all sets not themselves is and is not a member of itself. (If R is the set of all sets which do not contain themselves, does R contain itself? If it does then it does not and vice versa.) However, Cantor's theory of the infinite has survived controversy through the boldness of its idea as well as an almost cosmological beauty. As David Hilbert, the distinguished German mathematician famously remarked: 'No one can drive us from the Paradise Cantor has created for us.'[15]

This mathematics of the transfinite constitutes an alien world where the addition of a number to an infinite number leaves the latter unchanged. But what interested Matte Blanco was the fact that infinites have a unique quality: whole and part become equivalent. It should also be noted that the English psychoanalyst Wilfred Bion also linked infinity and emotion, although for him, infinity is Milton's 'formless' infinite for which there are no words.[16] This reminds us of MacNeice's 'sea which never thinks'.[17] It could be said, however, that both these poets attempt an approach to the infinite through language. In the philosophy of mathematics there has been a focus on whether infinite processes can be completed or not. Aristotle thought that there was only a potential infinite – that is, that we can count and count without ever reaching infinite completion. However, Plato thought that in some region outside space and time infinite counting processes could 'reach infinity' or achieve 'closure'.

Bilogic is of the opinion that consciousness, more at home with the finite, thinks of infinity as *potentially* infinite – that by endless counting, one is always 'getting there', whereas the unconscious point of view is that one could somehow complete infinite processes. The idea, originally due to Jung, of an unconscious 'imago' echoes this notion of a completed infinity. Unlike mental images, imagoes are seen as lacking nothing – totally terrifying or totally good. In this sense of super completeness, Matte Blanco thought, they correspond with infinite classes. This is effectively what Matte Blanco does in saying that, for example, the anxiety-making mother evokes the infinite class of mothers – this is really to say that the infinite uncontrollable mother imago is evoked. If someone is idealized and therefore 'all good', others must become 'all bad' and be deported into a split-off infinite class of bad things. This suggests the timeless quality of a Jungian archetype and here we find not only archetype as infinity but also negative infinity referred to in Bilogic.

15 See Constance Reid, *Hilbert* (London, 1970), p. 87. **16** Bion quotes John Milton's lines from book III of *Paradise Lost*: 'The rising World of waters dark and deep, / Won from the void and formless Infinite!' See Bion, *Transformations* (London, 1965), p. 151. **17** 'The Death-Wish', *Collected poems*, p. 199.

In various poems, including 'Littoral', MacNeice uses the image of the shore as a kind of between-world: between sea and shore, between infinite, spaceless and timeless core and finite world of the everyday. Thus the shore represents a kind of stage for 'longing back and aspiring forward' like the two-way movement of the tides. It must be said that MacNeice's 'longing back' is particularly intense, reminiscent of the title of Giorgio de Chirico's famous painting *Nostalgia of the Infinite* (1913). MacNeice's nostalgia could be summed up in a line from the section called 'Child's Terror' in his early work *Blind Fireworks* (1929): 'I am not; I have been; always the perfect tense'.[18] A life far from perfect is seen as perfect but in the past, in 'the perfect tense'; like completed infinities, the imago of home may lack for nothing in the imagination. But there is a paradox here, for there is a strong feeling that if MacNeice were to return home it would be like the return of the moth to the flame – it would be a return to death.

III

I would like to turn to the sea and infinity as 'other' and illustrate some themes so far only sketched or suggested. MacNeice describes the sea in 'Nature Notes' as being 'Incorrigible, ruthless,' and 'capable / Any time at all of proclaiming eternity';[19] in the last verse of 'Littoral' we find it addressed as that

> To whom we are ciphers, creatures to ignore,
> We poach from you what images we can,
> Luxuriously afraid
> To plump the Unknown in a bucket with a spade –
> Each child his own seashore.[20]

This near archetypal image of the sea is a good example of a medium – like the unconscious – not subject to conscious needs and wishes of human beings. MacNeice says in 'Experiences with images': '[the sea] was something alien, foreboding, dangerous and only very rarely blue'.[21] We also find 'unseen breakers' out in the dark somewhere and 'wolves of water who howl along our coasts'. Clearly the sea is perceived as infinite, wild and dangerous. MacNeice often uses sea images in quite a conventional manner: the sea-mother as underwater womb, drawing us down in a desire to 'possess the unposessable sea / As a man in spring desires to die in a woman.' As we sink below the surface of the sea we find ourselves in the symmetric 'heart of the Milky Way' where 'all our thinking gurgles down / Into the deep sea that never thinks'.[22] Here then is the symmetric core of all of us: spaceless and timeless, it resembles Jung's 'collective unconscious' in that

18 Ibid., p. 616. **19** Ibid., p. 549. **20** Ibid., p. 260. **21** *Selected prose*, p. 158. **22** *Collected poems*, pp 199–200.

not only difference has disappeared but the distinction between one human being and another.

But let us not neglect the surface of the poetry, for MacNeice liked diversity and difference, or in our terms asymmetry. In a later poem, 'Nature Notes' each section starts with the word 'Incorrigible' echoing the 'incorrigibly plural' of 'Snow'.[23] The word 'incorrigible' usually denotes a bad habit or person who cannot be changed. According to this poem, dandelions are incorrigibly extraverted and can grow anywhere. Incorrigible cats, incorrigible corncrakes – the world is 'incorrigibly plural'. Nature itself, MacNeice seems to suggest, contains a drive to difference or diversity; nature cannot 'help being this way'. Conscious Man sees the world as diverse not just in the sense of discrete facts but more in the nature of what W.B. Yeats called an 'uncomposite blessedness' in his poem 'Michael Robartes and the Dancer' – an image that suggests the presence of the symmetric behind the asymmetric.[24]

Now that we have isolated the domain of symmetric being – spaceless, timeless and infinite – we can finally turn to its logic. An extreme example from a psychotic patient can illuminate this form of thinking. This patient claimed he was President of Ireland on the grounds that both he and the President were Irish. He reasoned thus: the President is Irish; I am Irish; so I am the President. This so-called 'predicate thinking' is the leading idea of bilogic for, just as in conscious everyday life we classify people and things into groups, so also the unconscious mind classifies as well. The difference is this: consciously we might put elephants and computers in different classes but un-consciously (possibly revealed in a dream) they might be classified together as 'not forgetting'. Our patient above was dreaming in broad daylight. Consciously, a deserted house with no windows may have no obvious connection with death, yet unconsciously be perceived as a skull. Moving to a more sophisticated level, favoured by practitioners of Lacanian psychoanalysis, someone who dreams of having a steak, may reveal by free association that they were preoccupied by 'having a stake' in something.

MacNeice uses this (auditory rather than visual) similarity or predicate thinking a lot in his work. For example, in 'Eclogue by a Five-Barred Gate', Death speaks:

> D. I thought he was a poet and could quote the prices
> Of significant living and decent dying, could lay the rails level
> on the sleepers
> To carry the powerful train of abstruse thought –
> 1. What an idea!
> 2. But certainly poets are sleepers,
> The sleeping beauty behind the many-coloured hedge –[25]

23 Ibid., pp 548–49. **24** W.B. Yeats, *Collected poems* (London, 1990), p. 198. **25** *Collected poems*, pp 10–14.

The same sound of two kinds of 'sleepers' leads us from the railway to a 'powerful train of abstruse thought' whose sure foundation is poetry written by dreaming sleepers. On a lighter note, there is an apocryphal story told about Belfast at the height of the 'Troubles'. A visitor to the city who had wandered off the main tourist track was rendered almost speechless with fear when he was grabbed on the street and bundled into the back of a car. A gun was put to his head and a voice asked: 'Are you a Prod or a Taig?' 'I'm a Jew!' gasped the unfortunate man. Back came another question: 'Are you a Protestant Jew or a Catholic Jew?' This could be seen as a bilogical joke for, if we look at the story, first the tourist is asked if he is a member of the class of Protestants or the class of Catholics. When he replies that he is a member of neither class – the questioner (unconsciously) forms a larger class of Protestants, Catholics, Protestant Jews and Catholic Jews of which he presumably must be a member. This shows a kind of drive to abstraction in emotion.

On a more chilling note, in the process of isolating the Jews in Nazi Germany, Jewish people were made to wear a yellow star. The effect of this was to make them not only instantly recognizable in the street but it also pandered to some people's natural perceptual and emotional laziness, for at a glance they could think a person was just 'another Jew' if they saw her or him branded with the yellow star. So the Jews became as alike, in the minds of many, as ball bearings. The yellow stars encouraged others to similarize (i.e. symmetrize) Jews because they could not easily be distinguished from any other model citizens. The stars also enabled the hostility to irradiate to referring to the Jews as rats (you've seen one rat – you've seen them all) and eventually to make them invisible as a class. This was made considerably easier because they had, in the popular mind, become identical. People who have been reduced to dots or ball bearings in the public mind no longer evoke feelings because they have lost all individual characteristics and their differences have been eliminated by disgust. The point is illustrated in the scene on the Ferris wheel in Carol Reed's *The Third Man* (1949), where Harry Lime, played by Orson Welles, points to the people he sees as dots on the ground from high up on a Ferris wheel and asks who would not shoot a 'dot' for large financial reward. The antithesis to this pathology of people as indistinguishable and exterminable 'dots' is the numinous quality of individuality and uniqueness articulated in MacNeice's best-known poem, 'Snow':

> The room was suddenly rich and the great bay-window was
> Spawning snow and pink roses against it
> Soundlessly collateral and incompatible:
> World is suddener than we fancy it.
>
> World is crazier and more of it than we think,
> Incorrigibly plural. I peel and portion

> A tangerine and spit the pips and feel
> The drunkenness of things being various[26]

Here are the 'lots of lovely particulars' mentioned above, when MacNeice criticized his philosophy background in Oxford.

But to return to similarity thinking: the image of thread appears and reappears in MacNeice's poetry. Analogously, wool, string, cord, wire and even rope occur and recur, from first poems to last. At this point it is worth mentioning a remarkable case of the psychoanalyst Donald Winnicott. He was approached for a consultation: a mother was very concerned when her son persisted in tying the furniture together with string. Winnicott was able to intervene in the situation by pointing out that the boy was trying to symbolize a feeling of separation from his mother by this device. The use of string was a means of 'holding everything together'.[27] Similarly MacNeice's strings, threads and wires, perform this function in linking disparate things together. In one of his earlier poems, 'Reminiscences of Infancy', we find:

> Trains came threading quietly through my dozing childhood,
> Gentle murmurs nosing through a summer quietude,
> Drawing in and out, in and out, their smoky ribbons [...][28]

Clearly trains connect one place with another, and here the poet remembers a childhood train ride, sitting on his father's knee, from Belfast to his home at Carrickfergus.[29] In 'Morning Sun' (1935) the suggestion of weaving (in 'shuttle') makes a veritable tapestry of connections: 'Shuttles of trains going north, going south, drawing threads of blue ...'[30] But perhaps more obvious is the poet's connecting use of telephone 'wire' in Canto XXVI of *Autumn Sequel* (1953):

> The telegraph poles
> Endlessly filing past against the blue
>
> Remind me that those wires connect with living souls,
> Some few of whom I know [...][31]

Wire, rather than thread, links people by telecommunication. This question of linking and separation preoccupied MacNeice all his creative life. In the earlier poem, 'Cradle Song for Miriam', we find another variation on thread in the word 'strand': 'Till they gather the strands of us together / And wind us up for ever and

26 *Collected poems*, p. 24. **27** See Donald Winnicott, *Playing and reality* (London, 1971). **28** *Collected poems*, p. 615. **29** MacNeice, *The Strings Are False*, p. 38. **30** *Collected poems*, p. 15. **31** Ibid., p. 488.

ever.'[32] In the poem 'The Strand' another subversion occurs, linking the 'strand' or thread of life with the strand of the shoreline where MacNeice talks of following in the footsteps of his father paddling: 'my steps repeat // Someone's who now has left such strands for good'.[33] But it is in Canto V of *Autumn Sequel* that thread and shoreline are most fully intertwined: 'The wind braids / Long strands of brine together'.[34] Here, we have moved from thread of life to strand or shore, and thence to the sea as a source of all life. We see there are many different strands, threads and wires – not to mention ropes and cables in the poems – all of which are emotionally charged images for MacNeice. Each one represents the power of an infinite class of connections – an imago of connectedness, spaceless and timeless. This infinite notion dominates the poet's symmetrical core.

Finally, a note on the littoral – that space between low and high tides. Significantly, Winnicott regarded the shoreline as a symbol of the transitional space in which the young child comes to know the external world. That is, between the infinite sea and the finite land is the infinite-finite shore, MacNeice's 'between world'. This infinite-finite world represents the overlap between the subjective world and the world of other people and things for MacNeice. In 'Littoral' he links the shore with playing, which is reminiscent of Tagore's lines: 'On the seashores of endless worlds, children play'.[35] MacNeice also embodies playing in the game of 'cat's cradle', a string connecting game. The shore becomes a mental picture of the twilight state of mixed thoughts and feelings (feeling-thought) often referred to by writers on creativity. Unusually, MacNeice wonders about playing and whether sport might not be superior to art. The littoral as strand suggests a connection with the mother at the deepest level where land meets sea. The strand of life gathers up all the threads, wires, cables, strings, ropes into infinite connection at a very deep level of generalization and symmetric being. At this level, according to Matte Blanco, the differences between things disappear and in particular differences between people vanish too. This process is suggested in Canto XXVI of *Autumn Sequel* where MacNeice describes those who

aspire

To something more than selfhood; let the wall
Of isolation crumble and the light
Break in, but also out, the black scales fall

From all their eyes together in one white
And final annunciation.[36]

32 Ibid., p. 660. **33** Ibid., p. 263. **34** Ibid., p. 394. **35** Quoted by Winnicott in 'The location of cultural experience', *International Journal of Psychoanalysis* 48 (1967), 368–72. **36** *Collected poems*, p. 489.

In conclusion, I have attempted to show here how Matte Blanco's theory of 'bilogic' helps us see that MacNeice's poetry has an infinite core – spaceless, timeless and above all changeless. His preoccupation with difference in the poems is skin deep. But this is not to minimize it, for asymmetric difference interacts with symmetric similarity, and these logical strands, though solitary, do briefly fuse, in, for example, the 'drunkenness of things being various.' Matte Blanco, while stressing the independent nature of the two strands also allows contact in those moments he might call 'numinous'. But when MacNeice has another of his 'timeless' moments we are aware that the core of symmetric being has been touched, for example in the following lines from 'Meeting Point':

> God or whatever means the Good
> Be praised that time can stop like this,
> That what the heart has understood
> Can verify in the body's peace
> God or whatever means the Good.[37]

37 Ibid., p. 184.

'I'm an experiment': reflections on science in American short fiction

PHILIP COLEMAN

'The heart hid still in the dark, hard as the Philosopher's Stone.'[1]

In this essay I want to explore some of the ways in which American short story writers have engaged with science in their fictions. As a student of English literature and philosophy in University College Cork in the early 1990s, my only real contact with 'science' – apart from lectures in the Dairy Science building and occasional meetings with science students in the College bar – was a course on the philosophy of science I took in my second or third year. The central course text was Carl G. Hempel's classic if now somewhat outdated book *Philosophy of natural science*. First published in 1966, Hempel states in the book's opening chapter that:

> The high prestige that science enjoys today is no doubt attributable in large measure to the striking successes and the rapidly expanding reach of its applications. Many branches of empirical science have come to provide a basis for associated technologies, which put the results of scientific inquiry to practical use and which in turn often furnish pure or basic research with new data, new problems, and new tools for investigation.[2]

Hempel's book may have been replaced in recent years by more current writings in the philosophy of science – by figures such as Hilary Putnam and Lawrence Sklar – but his general point about the impact scientific and technological developments have on everyday life is perhaps even more relevant in the early twenty-first century than it was in the 1960s. However, as a student it struck me that the 'new data, new problems, and new tools for investigation' mentioned by Hempel are not only relevant to scientists and philosophers of science. Since the earliest

1 Paul Celan, *Collected prose*, trans. Rosmarie Waldrop (Manchester, 1986), p. 11. 2 Carl G. Hempel, *Philosophy of natural science* (Englewood Cliffs, NJ, 1966), p. 2.

times, creative artists – poets, story-tellers, painters, sculptors, and musicians – have endeavoured to understand the ways that new technologies and scientific discoveries alter our sense of the world. Moreover, the idea of human subjectivity itself has on numerous occasions been defined and redefined by science and scientific research. Hempel claims that science, 'apart from aiding man in his search for control over his environment [...] answers another, disinterested, but no less deep and persistent urge: namely, his desire to gain ever wider knowledge and ever deeper understanding of the world in which he finds himself.'[3] This is undoubtedly true, but the desire to gain a deeper understanding of the self is of particular importance for the artist in his or her engagements with scientific research and its objective reconfigurations of our sense of the world and how we live in it.

In my attempts to reconcile Hempel's ideas with my readings in English and American literature as an undergraduate I found the work of his contemporary Aldous Huxley, author of *Brave New World* (1932), briefly useful. Writing in 1963, Huxley attempted to describe the difference between 'science' and 'literature' in the following way:

> In the present context, science may be defined as a device for investigating, ordering and communicating the more public of human experiences. Less systematically, literature also deals with such public experiences. Its main concern, however, is with man's more private experiences, and with the interactions between the private worlds of sentient, self-conscious individuals and the public universes of 'objective reality', logic, social conventions and the accumulated information currently available.[4]

Huxley's statement, while it initially helped me to make what I thought was a necessary distinction between 'literature' and 'science', soon became problematic, however, not only because of its admission of contextual specificity and its rather vague delineation of 'public' and 'private' worlds, but also because it fails to acknowledge those writers whose work involves as much (and not 'less') systematic care in its composition as that taken by scientists in the preparation of laboratory experiments. Edgar Allan Poe, for example, was a scrupulously methodical author, and in many of his short stories he examines the influence of science not only on how the world is perceived but on the self who perceives it. In this regard he may be said to engage both the 'public' and the 'private' worlds described by Huxley, but Poe's insistence on the role of premeditated, objective care in the construction of literary texts is also at odds with his view of the artist as someone who operates at a certain remove from 'the public universes of "objective reality", logic, social conventions and the accumulated information currently available.'

3 Ibid., p. 8. 4 Aldous Huxley, *Literature and science* (London, 1963), pp 7–8.

David Van Leer has argued that the 'extravagance of [Poe's] narratives [has] encouraged readers to divorce [him] from intellectual and social issues and to imagine that he lived with his characters in some "ultimate dim Thule," a dream-land "out of SPACE – out of TIME".'[5] For many readers, the ghostly interiors of Poe's fictions and the sometimes violently surreal imaginings of his narrators would seem to reinforce Huxley's view of the writer as one primarily concerned with describing 'private worlds' and 'private experiences', but the construction of those fictional spaces was for Poe a matter of calculated analysis and experiment in the art of narrative form. As he put it in his celebrated review of Nathaniel Hawthorne's *Twice-Told Tales*, first published in May 1842:

> A skilful literary artist has constructed a tale. If wise, he has not fashioned his thoughts to accommodate his incidents; but having conceived, with deliberate care, a certain unique or *single effect* to be wrought out, he then invents such incidents, he then combines such events as may best aid him in establishing this preconceived effect. If his very initial sentence tend not to the outbringing of this effect, then he has failed in his first step. In the whole composition there should be no word written, of which the tendency, direct or indirect, is not to the one pre-established design. And by such means, with such care and skill, a picture is at length painted which leaves in the mind of him who contemplates it with a kindred art, a sense of the fullest satisfaction.[6]

Poe detailed with almost arithmetical precision the process by which his poem 'The Raven' was constructed in his essay 'The Philosophy of Composition',[7] but here, in what is widely regarded as a foundational text in the development of short story theory, he describes the extent to which the author must engage the full range of his analytical skills in the creation of a fictional 'tale'. It can be seen, in other words, that for Poe the 'literary artist' is a scrupulous technician who chooses words and constructs sentences with the same kind of meticulous attention to detail one might expect of a computer scientist writing a complicated program or a laboratory technician performing an important experiment – activities that are, like Poe's fictions, designed to achieve the 'outbringing' of a particular '*effect*'.

In literature as much as in science, of course, experiments do not always go according to plan, but Poe's view of the creation of literary works complicates Huxley's distinction between the writer's approach to the world and that of the scientist. Poe's interest in science is also fascinating, however, for the way that it challenges Hempel's understanding of the purpose of scientific inquiry. For Hempel,

5 David Van Leer, introduction, Edgar Allan Poe, *Selected tales* (Oxford, 1998), p. xxi. 6 Edgar Allan Poe, 'Nathaniel Hawthorne', *The selected writings of Edgar Allan Poe* (London, 2004), p. 692. 7 Ibid., pp 675–84.

science can assist human beings in their attempts to 'control' the environment, but it also satisfies our desire for what he calls an 'ever wider knowledge and ever deeper understanding of the world'. Poe's stories, as Van Leer suggests, are more concerned with the objective world than even the author himself liked to admit,[8] but they also provide insights into the ways that human subjectivity is radically reconceived at the same time that scientific research reconfigures our sense of the world in which we live. This aspect of science's influence – the way it shapes our sense of self – is overlooked by Hempel, and the extent to which literary artists have acknowledged science and embraced analytical methods in the history of literature is downplayed to a certain extent by Huxley, but in certain of Poe's short fictions we find these areas examined in considerable detail. In the work of Poe and his contemporary Hawthorne, indeed, the short story emerged in the nineteenth century as one of the most effective literary forms for examining the relationship between science and the self and, as we shall see, their examinations of the ethical and existential consequences of scientific research have provided important examples for a number of contemporary American writers of short fiction.

II

Bryant Magnum has written that F. Scott Fitzgerald used his early short stories 'as a workshop for subjects, themes, and techniques that he would continue to develop in his later stories and novels.'[9] Fitzgerald's early stories, in other words, may be seen as experiments, trials conducted before he would embark on the fuller, more adventurous projects that constitute the longer, more sustained works that are his major novels.[10] To take this view, however, is to underestimate the success of many of Fitzgerald's short stories in their own right, from early hits such as 'Winter Dreams' and 'Bernice Bobs Her Hair' to the late masterpieces 'Babylon Revisited' and 'The Lost Decade'. None of these stories are *about* science, and Fitzgerald rarely engages the subject of science in an explicit way in his fictions, long or short,[11] but a moment in his early story 'Head and Shoulders' provides a useful image (and the first part of the title) for this essay's examination of the place of science in American short fiction. In this story Fitzgerald describes what

8 Poe condemns what he calls 'the heresy of *The Didactic*' in his essay 'The Poetic Principle'. Ibid., pp 698–704. **9** Bryant Magnum, 'The short stories of F. Scott Fitzgerald' in Ruth Prigozy (ed.), *The Cambridge companion to F. Scott Fitzgerald* (Cambridge, 2002), p. 67 **10** Fitzgerald regarded his early story 'Absolution' as a 'prologue' to *The Great Gatsby*. See Magnum, p. 68. **11** Peter Freese, intriguingly, has read *The Great Gatsby* as a text that 'can be understood as an early fictional comment on the relentless workings of the Second Law [of Thermodynamics]'. See Freese, 'From the apocalyptic to the entropic end: from hope to despair to new hope?' in Peter Freese and Charles B. Harris (eds), *The holodeck in the garden: science and technology in contemporary American fiction* (Normal, IL, 2006), pp 334–56. The essays in Freese and Harris's collection focus almost exclusively on longer works of fiction, and only passing reference is made to the short story form.

happens when a brilliant young scholar at Princeton University named Horace Tarbox falls in love with a popular music hall singer called Marcia Meadow. Encouraged by the promise of 'five thousand Pall Malls if she would pay a call on [this] prodigy',[12] Marcia visits Horace in his flat to find him contemplating modern philosophy. 'I am a realist of the School of Anton Laurier – with Bergsonian trimmings – and I'll be eighteen years old in two months', he declares, to which Marcia replies: 'Well, that's O.K. with me. I got a notion I want to see you do something that isn't in your highbrow programme. I want to see if a whatch-call-em with Brazillian trimmings – that thing you said you were – can be a little human.'[13]

In the course of Fitzgerald's story Tarbox becomes 'human' by falling in love with Marcia, marrying her, and giving up his sedentary life as a scholar, but before this can happen he has to opt out of the 'experiment' that has been his life up to this point. As he explains to Marcia when they first meet:

> You see I'm a plan. I'm an experiment. I don't say that I don't get tired of it sometimes – I do. [...] Here's my history: I was a 'why' child. I wanted to see the wheels go round. My father was a young economics professor at Princeton. He brought me up on the system of answering every question I asked him to the best of his ability. My response to that gave him the idea of making an experiment in precocity. To aid in the massacre I had ear trouble – seven operations between the ages of nine and twelve. Of course this kept me apart from other boys and made me ripe for forcing. Anyway, while my generation was laboring through Uncle Remus I was honestly enjoying Catullus in the original.[14]

The idea of the self as 'an experiment' is something that Fitzgerald may have inherited from his American precursors Hawthorne and Poe, each of whom saw the relationship between scientific research and its impact on the self in decidedly less comical terms than Fitzgerald sees it here. In their work, indeed, men and women are frequently encountered as the subjects of experiments conducted by figures who are said to represent various aspects of the world of science and scientific thought, the most (in)famous of whom is probably Poe's Monsieur C. Auguste Dupin.

The best-known representations of nineteenth-century scientific theory in Poe's work are his Dupin tales and *Eureka*, his 'essay on the material and spiritual universe' which is often said to prefigure twentieth-century ideas in quantum theory in some of its claims about the nature of space and time. I want to focus here on 'Von Kempelen and His Discovery', a piece first published in April 1849,

12 F. Scott Fitzgerald, *Novels and stories 1920–1922* (New York, 2000), p. 311. **13** Ibid., p. 316. **14** Ibid., p. 315.

only a few months before Poe's death, in which medieval and modern ideas about 'science' are brought into play and where Poe poses questions about scientific research that are as relevant today as they were during his lifetime. Poe was fascinated by the Hungarian scientist Johann Wolfgang von Kempelen (1734–1804), whose automated chess-player was exhibited in the United States by J.N. Mäelzel in the 1830s and became the subject of his story 'Maelzel's Chess-Player'. Poe's satirical send-up of the California Gold Rush in 'Von Kempelen and His Discovery' plays on the idea that von Kempelen also discovered a method for turning lead into gold.[15] Poe's piece manipulates various literary forms – the journalistic essay, the fictional tale, and the scholarly article – in a manner that could be said to replicate the alchemical processes the narrator claims were employed by von Kempelen in his pursuit of what medieval scholars called the *lapis philosophorum* or 'Philosopher's Stone.' Manipulating these various literary forms, however, Poe creates a text that not only tells us very little about von Kempelen's so-called 'discovery', but one which deliberately exposes both the debasement of knowledge by the desire for monetary gain and the disintegration of the self at the very moment when the pursuit of intellectual wisdom seems to have achieved its aim. While the text treats a scientific topic, then, Poe is more concerned with analysing the impact of science on human subjectivity – for scientists and non-scientists alike – than he is with exploring the details of scientific theory in themselves.

Poe shifts the reader's attention from purely scientific matters – what he calls, with curious but characteristic emphasis, the '*scientific* point of view' – to the issue of 'von Kempelen himself' – in the opening paragraph of the story:

> After the very minute and elaborate paper by Arago, to say nothing of the summary in 'Silliman's Journal', with the detailed statement just published by Lieutenant Maury, it will not be supposed, of course, that in offering a few hurried remarks in reference to Von Kempelen's discovery, I have any design to look at the subject in a *scientific* point of view. My object is simply, in the first place, to say a few words of Von Kempelen himself (with whom, some years ago, I had the honor of a slight personal acquaintance,) since everything which concerns him must necessarily, at this moment, be of interest; and in the second place, to look in a general way, and speculatively, at the *results* of the discovery.[16]

The relegation of 'the *results* of the discovery' to second-place in the narrative, after a consideration of 'Von Kempelen himself', is important because it signals the extent to which Poe was aware of what is today a frequent hindrance in public

15 Poe's texts refer to the Hungarian scientist incorrectly as 'Von Kempelen.' I refer to him here as 'von Kempelen.' For further details about von Kempelen's life and work see Tom Standage, *The Turk: the life and times of the famous eighteenth-century chess-playing machine* (New York, 2002). **16** Poe, *Selected tales*, p. 319.

debate about scientific discoveries: namely, the celebritization of scientific personalities, which often means that more media time is spent discussing the private life of Stephen Hawking, for example, than his theories of the universe.[17] Poe's narrator is aware of the public interest in the lives of scientists, especially those whose discoveries present major challenges to our sense of self, and this is one of the reasons why he considers 'von Kempelen himself' before he discusses the particulars of his scientific research. As with much contemporary sensationalist journalism – a form of popular literary production with which Poe was very familiar[18] – his narrator claims to have *the* story on the scientist in question. 'In the brief account of Von Kempelen which appeared in the "Home Journal",' he writes, 'several misapprehensions of the German original seem to have been made by the translator, who professes to have taken the passage from a late number of the Presburg "Schnellpost".' He continues:

> 'The Literary World' speaks of him, confidently, as a *native* of Presburg (misled, perhaps, by the account of him in the 'Home Journal',) but I am pleased in being able to state *positively*, since I have it from his own lips, that he was born in Utica, in the State of New York, although both his parents, I believe are of Presburg descent.[19]

The frequent italicizations and assertions are intended to present the narrator not only as a confident if somewhat self-assured authority on 'Von Kempelen's discovery', but on von Kempelen 'himself', who is presented here as an American. This, of course, makes him seem much closer, in social and cultural terms, to Poe's contemporary readers than his presentation as a European would have meant. It could be said, in other words, that Poe reinvents von Kempelen as an American because he wants to suggest to his readers that he is one of their own.

In his presentation of the character of 'Von Kempelen' in this piece, then, Poe deliberately distorts the reader's sense of the relationship between the fictional and the real. His presentation of him as a living American personality when in fact the 'real' von Kempelen was a Hungarian who died over four decades before Poe's piece was published is clear evidence of the author's manipulation of 'truth' in this narrative but, as we have seen, the narrator of the text is extremely concerned to be seen as the one who can '*positively*' reveal the truth about 'Von Kempelen'. As he writes at one point, however, 'the truth *may be* stranger than fiction'.[20] In the history of science the truth of this claim has often been demonstrated: scientific

17 For an interesting account of the celebritization of scientists including Richard Dawkins, Stephen Jay Gould, Stephen Hawking, Carl Sagan, Steven Weinberg, and Edward O. Wilson, see Karl Giberson and Mariano Artigas, *Oracles of science: celebrity scientists versus God and religion* (Oxford, 2007). **18** See G.R. Thomson, 'Popular fiction: *Blackwood's* and the sensation tale' in Poe, *The selected writings of Edgar Allan Poe* (New York and London, 2004), pp 754–5. **19** Poe, *Selected tales*, p. 321. **20** Ibid., p. 322.

discoveries frequently reveal things to us that have previously been thought impossible or improbable, from the discovery that the Earth is not the centre of the universe to the detection of sub-atomic particles. These discoveries not only altered our sense of the world of things but they also changed our sense of who we are, and this is the central point of Poe's story. Von Kempelen, as a direct consequence of his discovery of a method for turning lead into gold, is suspected of being a criminal, because of his sudden and unaccountable acquisition of great wealth. Changes in the way the scientist is perceived, however, give way in the course of the story to a description of the narrator's sense of how society will change as a result of his discovery, which Poe links to the California Gold Rush, an event which was of intense public interest on both sides of the Atlantic in the late 1840s:

> Speculation, of course, is busy as to the immediate and ultimate results of this discovery – a discovery which few thinking persons will hesitate in referring to an increased interest in the matter of gold generally, by the late developments in California; and this reflection brings us inevitably to another – the exceeding inopportuneness of Von Kempelen's analysis.[21]

For good or ill, Poe suggests, scientific discoveries have consequences far beyond the laboratories in which they are made, not least in terms of their ability to alter the course of individual – and communal – fate.

On one level, then, 'Von Kempelen and His Discovery' describes in a deliberately exaggerated and sensationalist fashion the impact that scientific research can have on the way people think and behave. Moreover, it reveals how quickly a 'scientific discovery' – perhaps even more so than 'discoveries' in other forms of intellectual and creative inquiry – can overturn our sense of the world and our place in it. Poe's story is of course a complete fabrication, designed in part to dupe his nineteenth-century readers, but it does nonetheless speak to the issue of science's rhetorical or discursive power in affecting individual and communal consciousness. Before the story concludes, Poe's narrator states that: 'It is, indeed, exceedingly difficult to speculate prospectively upon the consequences of the discovery', punning a little heavy-handedly on the fate of the many gold prospectors who are described in the text's penultimate paragraph. However, his concern is not just with the economic fate of those foolhardy men. The story's final paragraph reads: 'In Europe, as yet, the most notable results have been a rise of two hundred per cent. [*sic*] in the price of lead, and nearly twenty-five per cent. [*sic*] in that of silver.' These fictional fiscal figures are not as revealing as the clause 'as yet' at the start of this sentence, by which Poe suggests that the *actual* 'consequence of the

21 Ibid., pp 324–5.

discovery' of von Kempelen may not be calculable in the same way that stockbrokers can estimate the relative market values of precious metals. Changes in individual and cultural identity, Poe suggests, are not as easily measured as fluctuations in the value of stocks and shares. His strange account of von Kempelen invites us to consider the ramifications of scientific research which extend, he claims, beyond the realm of verifiable fact and empirical observation.

The fictional nature of the 'discovery' described in Poe's story does not lessen its importance as a text that explores the relationship between scientific research and social change, then, and from the description it provides of the changes wrought by his findings on von Kempelen himself to the effects his 'discovery' will have on individuals around the globe, 'Von Kempelen and His Discovery' explores many issues relating to the place of science in human affairs during the first half of the nineteenth century. At a time when the scientific speculations and discoveries of Luigi Galvani, Sir William Herschel, Sir Humphrey Davy, Sir Charles Lyell, and others,[22] were presenting a profound challenge to traditional understandings of the idea of the 'human', it is no surprise that Poe should have sought to explore certain aspects of 'science' in his fiction, and it is certainly something that can be remarked in the work of other short fiction writers of the period, and Poe's contemporary Nathaniel Hawthorne (1809–64) in particular. In stories like 'The Birthmark', 'Dr Heidegger's Experiment' and, most notably, 'Rappaccini's Daughter', Hawthorne examines the problem of the consequences of scientific inquiry in ways that are arguably even more unsettling than anything encountered in the work of Poe.

Possibly the most famous 'scientist' in Hawthorne's work is Roger Chillingworth, the first husband of Hester Prynne and the man who makes a special (one could say 'scientific') study of Arthur Dimmesdale, the Puritan minister with whom Hester has a child (Pearl) in *The Scarlet Letter* (1850). Dimmesdale is the subject of Chillingworth's experiment, and early in the narrative we are told that he (Chillingworth) is 'a man of skill in all Christian modes of physical science' as well as 'whatever the savage people could teach, in respect to medicinal herbs and roots that grew in the forest.'[23] Set in seventeenth-century New England, *The Scarlet Letter* is a book that also says a great deal about the nineteenth-century contexts within which it was written. Moreover, its depictions of the scientist may be said to address contemporary fears about the impact that scientific discoveries were having on religious faith and ideas about the primacy of human being in the world. At one point in the narrative, in a description of

22 See Luigi Galvani, from *De viribus electricitatis* (1791), Sir William Herschel, from *On the power of penetrating into space by telescopes* (1800), Sir Humphrey Davy, from *Discourse, introductory to a course of lectures on chemistry* (1802), and Sir Charles Lyell, from *Principles of geology* (1830–3), in Laura Otis (ed.), *Literature and science in the nineteenth century: an anthology* (Oxford, 2002), pp 135–40, 43–7, 140–43, 246–52. **23** Nathaniel Hawthorne, *The Scarlet Letter* (Oxford, 1990), p. 70.

Hester Prynne designed to provide 'another view' of protagonist, we are told that she lived in 'an age in which the human intellect, newly emancipated, had taken a more active and a wider range than of many centuries before.'[24] This description applies as much to the nineteenth century as it does to the seventeenth-century world in which Hester is forced to wear a scarlet letter for her sins, but Hawthorne is careful, here and elsewhere in his work, to point out the dangers and the limitations of such intellectual emancipation. These are summarized by Dimmesdale in a scene where he – as the unwitting subject of an experiment – speaks back to Chillingworth, the scientist, telling him that there are certain kinds of knowledge that will always remain just beyond his reach. The minister has asked the physician where he obtained certain 'unsightly plants' – indicating his keen eye for one particular branch of science, botany – and he suggests that they have grown out of a grave in the cemetery, as if to 'typify […] some hideous secret' that was buried with the man who lay there. This is hardly a very 'scientific' claim on Chillingworth's part, but to the suggestion that 'all the powers of nature call […] for the confession of sin' – that there is, in other words, an inherent connection between the things of the world and the self, and that these connections can be discerned and revealed by scientific research – Dimmesdale replies:

> There can be, if I forebode aright, no power, short of Divine mercy, to disclose, whether by uttered words, or by type or emblem, the secrets that may be buried with a human heart. The heart, making itself guilty of such secrets, must perforce hold them, until the day when all hidden things shall be revealed. Nor have I read or interpreted Holy Writ, as to understand that the disclosure of human thoughts and deeds, then to be made, is intended as a part of the retribution. That, surely, were a shallow view of it. No; these revelations, unless I greatly err, are meant merely to promote the intellectual satisfaction of all intelligent beings, who will stand waiting, on that day, to see the dark problem of this life made plain.[25]

The desire to see 'the dark problem of this life made plain' is of course something that both Dimmesdale and Chillingworth share, though the minister is willing to believe that some things will remain a mystery until what he calls the 'last day' of divine revelation.

Hawthorne's text exposes the divide between scientific and religious understanding that was becoming increasingly apparent in the early decades of the nineteenth century, and the debate is played out here and in other writings by him as an existential drama where the meaning of human subjectivity itself is the

24 Ibid., p. 164. **25** Ibid., pp 131–2.

central focus. In his short story 'Rapaccini's Daughter', first published in the *Democratic Review* in December 1844, the issue is explored with characteristic insight and power. Here we are presented with a character (Dr Giacomo Rappaccini) who is said to have 'as much science as any member of the faculty – with perhaps one single exception – in Padua, or all Italy', but who also 'cares infinitely more for science than for mankind'.[26] Such is his devotion to science, indeed, that '[h]is patients are interesting to him only as subjects for some new experiment.' In the course of the story we learn how Rappaccini uses his own daughter in an experiment to create an antidote to all known poisons, partly because of his desire to protect her from the world, but also to further his own scientific interests: 'with what he calls the interest of science before his eyes,' we are told, Rappaccini 'will stop at nothing.'[27] The story centres on the infatuation of a young student (Giovanni Guasconti) with the scientist's daughter (Beatrice), but Hawthorne's description of their doomed relationship is less interesting for its portrayal of their courtship than it is for the way it explores Giovanni's pursuit of a scientific understanding of the experiment in which he also plays a crucial part. His method may be to work with materials which he claims are 'the most opposite to those by which [Beatrice's] awful father has brought this awful calamity' upon them, but like her father he is also a scientist, and at the end of the story she dies 'at the feet of her father *and* Giovanni.'[28]

'Rappaccini's Daughter', then, invites us to consider the ways that science can transform the self, but Hawthorne leaves it to the reader to ponder the ethical implications of scientific research in general. The story ends with another scientist, Professor Pietro Baglioni, calling 'in a tone of triumph mixed with horror, to the thunder-stricken man of science, – "Rappaccini! Rappaccini! and is *this* the upshot of your experiment?"'[29] Baglioni assumes a morally superior position here and elsewhere in the story – he addresses Rappaccini from an upstairs window overlooking the garden where Beatrice has died – but the 'triumph' in his tone of voice suggests that his judgement is not only moral but it also has to do with a sense of competitiveness between the two old scientists. This is already suggested, indeed, by the phrase 'with perhaps one single exception' in Baglioni's description of Rappaccini's scientific pre-eminence earlier in the tale: the 'purity' of his ethical interest in the tragic fate of Rappaccini's daughter may be seen as secondary to more selfish, purely professional concerns.

'Rappaccini's Daughter' was first published in book form in 1846 in Hawthorne's collection *Mosses from an Old Manse*. Herman Melville, whose novel *Moby-Dick* offers extended treatments of another area of scientific inquiry – cetology or the branch of zoology that deals with whales and other aquatic mammals

26 Nathaniel Hawthorne, 'Rappaccini's Daughter' in James McIntosh (ed.), *Nathaniel Hawthorne's tales* (New York and London, 1987), p. 192. **27** Ibid. **28** Ibid., p. 209. Emphasis added. **29** Ibid. Emphasis in original.

– wrote an essay based on *Mosses from an Old Manse* in which he describes the 'wondrous symbolizing of the secret workings in men's souls' in Hawthorne's fictions. 'For spite of all the Indian-summer sunlight on the hither side of Hawthorne's soul,' he writes, 'the other side – like the dark half of the physical sphere – is shrouded in a blackness, ten times black.'[30] The characters of Roger Chillingworth and Giacomo Rapaccini may be said to represent the 'dark' sides of spheres that are completed when they are coupled with their ostensible opposites, Arthur Dimmesdale and Pietro Baglioni. As Melville makes clear, however, 'this darkness but gives more effect to the evermoving dawn, that forever advances through it, and circumnavigates his world.'[31] In Hawthorne's work science is presented as something that can and does advance understanding of the world and the self who inhabits it, but he seeks at every turn, like Poe, to remind his readers that science as an activity of the human mind also has its limits and limitations.

III

In Poe's 'Von Kempelen and His Discovery' and Hawthorne's 'Rappaccini's Daughter' the subject of scientific research leads both authors to consider questions that cannot be readily comprehended by empirical methods. The twentieth-century Czech scientist-poet Miroslav Holub often succumbed – if that is the word – to an abandonment of the purely rational intellect for the imaginative freedom afforded by the art of poetry in an attempt to understand similar kinds of problems. In his poem 'Heart Transplant', for example, he uses a series of similes to describe the interval in cardiothoracic surgery when the body is without its own heart, just before a new heart has been given to it, in the following terms:

> After an hour
>
> there's an abyss in the chest
> created by the missing heart
> like a model landscape
> where humans have grown extinct.
>
> [.............................]
>
> It's like falling from an aeroplane, the air growing cooler and cooler,
> until it condenses in the inevitable moonlight,
> the clouds coming closer, below the left foot, below the right foot,

30 Herman Melville, 'Hawthorne and his Mosses' in McIntosh (ed.), *Nathaniel Hawthorne's tales*, p. 341. **31** Ibid.

a microscopic landscape with roads and capillaries pulsing in counter-
movements,
feeble hands grasping for the King of Blood[32]

Here Holub admits that there are areas of experience beyond what we can observe where poetry seems more useful than science for exploring gaps in our understanding. The poem uses the literally heartless space of the chest in surgery as a figure for the absence of the poetic or imaginative in over-rationalized explanations of the world, but the re-placement of the heart into the body is described also in terms of a return to the possibilities of the poetic with the celebration of the 'King of Blood' who will allow 'Life and Spirit', as he says further on in the poem, to thrive as one:

Atrium is sewn to atrium,
aorta to aorta,
three hours of eternity
coming and going.

And when the heart begins to beat
and the curves jump
like synthetic sheep
on the green screen,
it's like a model of a battlefield
where Life and Spirit
have been fighting

and both have won.[33]

For Holub, in this poem at least, physical science ultimately coincides with the metaphysical propensities of poetry in a way that recognizes their mutual value for understanding and celebrating 'the human heart'. Holub's refusal to prioritize one form of understanding is important, moreover, because it suggests that both 'science' and 'literature' may be said to participate in the common pursuit of self-knowledge. Hawthorne and Poe, as we have seen, also recognized the usefulness of literature as a medium for exploring scientific claims about the nature of the world and our place in it, but they went further to suggest that scientific understandings of reality need to be complemented by the ethical and existential interrogations of the creative imagination.

Contemporary American writers such as David Foster Wallace and George Saunders have extended this tradition of inquiry into the twenty-first century, and

32 Miroslav Holub, *Vanishing Lung Syndrome*, trans. David Young and Dana Hábová (London, 1990), p. 55. **33** Ibid., p. 56.

the short story has been recognized by them as a profoundly effective form for examining ideas that are of general public concern. Wallace and Saunders are particularly interesting writers because they both have background training in the pure and applied sciences – Wallace studied logic and the philosophy of mathematics at Amherst College, while Saunders studied geophysical engineering at the Colorado School of Mines. Like Hawthorne and Poe, their work often tends towards the grotesque and the extreme, but like their nineteenth-century precursors they are also concerned with the question of the impact that scientific and technological developments have on the individual self. Two examples – one story by each author – will be sufficient to illustrate the powerful ways in which Wallace and Saunders have realized the usefulness of short fiction for exploring the role of science in contemporary life.

First published in the magazine *McSweeney's* in 2000 under the pseudonym Elizabeth Klemm, and subsequently printed as the first story in his 2004 collection *Oblivion*, Wallace's 'Mister Squishy' describes on one level the dynamics between a group of individuals who have been put together as part of a corporate product assessment exercise by the fictional company 'Reesemeyer Shannon Belt Advertising.'[34] 'Mister Squishy' is in fact a manufacturer of 'snack cakes' and the 'Focus Group' described in the story ('Team Δy') is assembled to test a new product – the 'dark and exceptionally dense and moist-looking […] *Felonies!*® – a risky and multivalent name meant to connote and to parody the modern health-conscious consumer's sense of vice/indulgence/transgression/sin vis à vis the consumption of a high-calorie corporate snack.'[35] In the course of the story Wallace describes in detail the crossover between scientific research and corporate finance, and he explores the extent to which marketing strategies are informed by developments in laboratories where the colours, textures, and flavours of 'snack cakes' are manufactured. Drawing on the worlds of finance (and especially the fields of advertising, human resources and marketing), as well as mathematical, behavioural and food science, Wallace provides a disturbing insight into the ways that research in these areas often seeks to alter our sense of who we are. As the story unfolds and details about various characters are revealed, it becomes clear that experiments to test *Felonies!*® and other products on 'Team Δy' have had a detrimental effect on some members of the group, and on one member in particular – Schmidt – who plans to poison the other participants.

The story represents on one level the complex economic, scientific, social and subjective processes that are involved in the creation of all marketable commodities, as well as the choices individual consumers make when they decide to purchase them, but its presentation of Schmidt as an evil (domestic) scientist provides an even more sinister twist to the insights the story provides into modern

34 David Foster Wallace, *Oblivion: stories* (London, 2004), p. 3. **35** Ibid., p. 5.

life. Schmidt turns 'the system' back on itself when he realizes that, as Wallace writes:

> [...] even though so many upmarket consumer products now were tamperproof, Mister Squishy – brand snack cakes – as well as Hostess, Little Debbie, Dolly Madison, the whole soft-confection industry with its flimsy neopolymerized wrappers and cheap thin cardboard Economy Size containers – were decidedly not tamperproof at all, that it would take nothing more than one thin-gauge hypodermic and 24 infinitesimal doses of KCN, As_2O_3, ricin, $C_{21}H_{22}O_2N_2$, acincetilcholine, botulinus, or even merely Tl or some other aqueous base-metal compound to bring almost an entire industry down on one supplicatory knee; for even if the soft-confection manufacturers survived the initial horror and managed to recover some measure of consumer trust, the relevant products' low price was an essential part of their established Market Appeal Matrix*, and the costs of reinforcing the Economy packaging or rendering the individual snack cakes visibly invulnerable to a thin-gauge hyperdermic would push the products out so far right on the demand curve that mass-market snacks would become economically and emotionally untenable, corporate soft confections going this the way of hitchhiking, unsupervised trick-or-treating, door-to-door sales, &c.
>
> * also, somewhat confusingly, = MAM[36]

Wallace's narrator seems to have a very thorough knowledge of modern food science but he also knows a thing or two about the dark arts practiced in the name of 'science' by Roger Chillingworth in Hawthorne's *The Scarlet Letter*. In the character of Schmidt, as in Chillingworth and, indeed, Rapaccini, we encounter a man who exposes the fine line between ethical and unethical scientific research. What makes Wallace's story particularly troubling, however, is not only the description of Schmidt's capacity for evil, but its exposition of the degree to which scientific research is used to deceive a public that has learned to trust and believe the word of 'MAM' – the caring face of corporate culture.

Wallace has been described as a 'superb comedian of culture',[37] but he is also a serious commentator on contemporary society, and one of the most important aspects of a story like 'Mr Squishy' is the way that it forces readers to think about issues such as the ethics and economics of scientific research. In his story '93990', from his 2006 collection *In Persuasion Nation*, George Saunders is similarly concerned to go behind laboratory doors, as it were, to describe the ways that science is conducted in the name of progress. The story begins as follows:

36 Ibid., pp 30–1. Wallace's habitual use of footnotes in his fiction is in evidence here. **37** 'Praise for David Foster Wallace', back cover of *Oblivion: stories* (London, 2004).

> A ten-day acute toxicity study was conducted using twenty male cynomolgus monkeys ranging in weight from 25 to 40 kg. These animals were divided into four groups of five monkeys each. Each of the four groups received a daily intravenous dose of Borazadine, delivered at a concentration of either 100, 250, 500, or 10,000 mg/kg/day.[38]

Like Wallace's 'Mr Squishy', Saunders' story draws on scientific terminology, but at no point in the story does a subjective narrator intervene to describe an emotional or affective response to the experiment being conducted. The story moves towards its concluding sentence – 'All carcasses were transported off-site by a certified medical waste hauler and disposed of via incineration'[39] – without any clear acknowledgement of human presence, apart from passing references to the monkeys' handlers and the decision, on the tenth day of the experiment, to 'increase [93990's] dosage to 100,000 mg/kg/day, a dosage which had proved almost immediately lethal to every other animal in the highest-dose group', which is 'adjudged to be scientifically defensible.'[40] The story not only focuses the reader's attention on what the monkeys are forced to endure, but it also closes the gap between the world of scientific research and the private world in which the products of that research are blithely consumed – from cosmetics to convenient foods. Saunders attempts to break down the barriers that exist between different forms of understanding and, like Wallace, what he exposes beneath the surface of things is often profoundly disturbing.

The short stories of David Foster Wallace and George Saunders discussed here bring us back to a consideration of what it means to be human in a world that, as Saunders suggests, is constructed by the arts *and* sciences of 'persuasion.' Given their relative brevity and the fact that they are often first published in magazines or journals with large and sometimes international readerships, short stories have often been highly effective in raising consciousness about a range of issues – from the construction of regional identity in the stories of Sarah Orne Jewett and Charles W. Chesnutt to the interrogation of racial politics in the short fiction of Richard Wright and Flannery O'Connor. Their popularity in the American context has been described by Dave Eggers, one of the form's most notable contemporary practitioners (and publishers), in his introduction to *The Best of McSweeney's*. There he writes that 'The US has long been short-story crazy,'[41] but it is important to realize that its popularity has allowed writers of short fiction to introduce large numbers of readers to issues that they might not otherwise encounter, including considerations of the ethical and existential consequences of modern scientific research. The purpose of this essay has been to explore some of

38 George Saunders, *In Persuasion Nation* (New York, 2006), p. 109. **39** Ibid., p. 117. **40** Ibid., pp 116–17. **41** David Eggers, 'Introduction,' *The best of McSweeney's*, vol. 1 (London, 2005), p. vii.

the ways that American writers of short stories – from Edgar Allan Poe and Nathaniel Hawthorne in the nineteenth century to Wallace and Saunders in the contemporary period – have interrogated the methods, findings, and discoveries of different kinds of science. In each case science is seen as a discourse that alters the way we see the world, but the authors concerned are also keen to expose the ways that developments in science challenge our sense of who we are. It might be objected that the writers discussed here take a generally negative view of science in their fictions, or that their representations of science and scientists bear no relation to real life. Reading them, however, we are encouraged to accept that neither the advances of science nor the perfections of art are capable of achieving the kind of absolute power sought after by characters such as Poe's Von Kempelen, Hawthorne's Rappaccini, Wallace's Schmidt or Saunders' unnamed laboratory assistant. As the Romanian poet Paul Celan wrote in one of his prose fragments: 'Do not be deceived: this last lamp does not give more light – the dark has only become more absorbed in itself.'[42]

42 Paul Celan, *Collected prose*, p. 14.

Can poetry be scientific?

PETER MIDDLETON

This is a question usually answered in the negative even by lovers of poetry, sometimes in a tone of dismissal – if science is now the most reliable means of understanding the world why should we look to poetry for anything other than personal expression of feeling? – and sometimes with defiance – poetry is needed to counter the mechanistic, materialist, devitalized scientific picture of the world. We rarely think that there might be a positive answer, that poetry might be capable of researches that augment or parallel those of science. I say 'we' and yet this is not quite accurate. Some modernist poets of each generation throughout the twentieth century have made such claims for their poetry, and tried to articulate the implications in their essays as well as to create poems that experiment, research and observe objectively what has not yet been understood about the world. These aspirations and achievements have not made much impression on literary studies because the criticism of poetry has been dominated for several decades by the view that poetry is infused with a subjectivity that dissolves reasoning and epistemology, and turns language into the music of an internally referential dance of signifiers. Added to the problem of a theoretical paradigm resistant to considering claims for truth and knowledge is a general lack of information about what constitutes contemporary science. What is mathematics, what goes on in a laboratory, and how certain or provisional is scientific knowledge? Literary historians and scholars mostly lack advanced scientific training.

I can't claim advanced expertise myself, though I did study mathematics at university before switching to English literature, but I do believe that even without specialist training in the sciences we can still offer more persuasively affirmative answers to my opening question. We have all grown up with both science and literature, and I suspect my own experiences of an early interest in science are not that dissimilar from those of many literary theorists of poetry. My own fascination with the insights emerging from the sciences and their implications for contemporary culture began during the two years I spent in Washington DC as a teenager in the nineteen-sixties. Our family was there because my father was working with the British Embassy and needed to live centrally in the city. We lived for two years in a small numbered street lined with trees just off Connecticut Avenue while Kennedy was shot, the threat of imminent nuclear attack slowly waned, the space race continued, and the pressures that would lead to the Watts riots grew. It had taken us nearly a week by ship to reach America. Surrounded by so much modern building and technology, my unheated, old-fashioned country where television was still a novelty seemed remote in time as well as far across the ocean, a distance marked out by scientific advance measured in cars, electronics, nuclear weaponry, and even school intercoms. Our neighbourhood of government employees was not a good area for kids; almost all the houses were childless, and I developed a keen interest in what Americans called 'hobbies', building radios and intercoms. At school I became fascinated by the evolutionary hierarchy from protozoa to chordates that we studied daily in biology. When I wasn't holding a soldering iron or listening in on other rooms in the house I was reading something called 'science fiction' I discovered in a special bookcase in the small public library round the corner. Soon we would all be driving air cars, flying to other planets, and trying out all sorts of new gadgets for communication or time travel, and I wanted to be in on it.

When I returned to England to a newly rebuilt Grammar School, now housed in a modern building that blended heritage style medieval open cloisters with the new modernist brutalism I discovered that I would not be able to reconcile arts and science in the same way. You had to make a decision between two sorts of A-levels, a choice that would then shape the rest of your life, or so it seemed. Not only did I like both mathematics and English – I was always reading several novels at once – it represented a choice between my parents. Both of them were war dropouts: my mother had a place to study history at university but went into the WRENS instead, though she continued to read histories and developed a keen interest in local archaeology; my father, who had endured a year of modern languages at university knowing that he should have chosen mathematics before being drafted into the Bletchley code-breaking effort, became a reader of science books and magazines (and eventually took an Open University degree in geology in his seventies after my mother finished hers in literature).

I opted for A Levels in double mathematics and physics, and took a second O level in English, hoping this would be the synthesis that would work. It didn't. I went to Oxford to study PPE thinking this was another possible integration of scientific method and narrative imagination applied to the study of social life, realized at once that it was a mistake, and spent my first year oscillating between mathematics lectures and tutorials and English lectures and tutorials, took examinations in mathematics and then switched to English. This was hard and it took me another two years to catch up, and to learn how to write an essay and accept the fuzzy logic of literary criticism after the objective clarities of science, always feeling that I was an impostor, someone without that necessary sixth-form training in English Literature. In the midst of my anxieties about this I once asked William Empson after a poetry reading if he had ever regretted his own abandonment of mathematics for literary criticism and he asked me to stop blowing cigarette smoke in his face, the question ignored. Gradually the question lost its grip on me too and science faded out of sight. By the time I was working on a doctorate, I felt that I had made the correct choice or rather forgot that it had ever felt like a choice. I have heard many similar stories from colleagues and writers.

In the past decade those earlier interests began to reawaken and I am not entirely sure why. Perhaps it was the diminishing hegemony of literary theory. An exposition of deconstruction which assumed that a binary opposition would always be a clearly demarcated division might trigger off memories of Hausdorf spaces from topology, where I seem to recall that although you had two independent sets of points you could never find the edge of either nor the boundary line between them. Did mathematics have images for complexity that literary theory could benefit from? Sometimes I would wonder why literary scholars could be so exacting in their attentions to one theory of mind or language and indifferent to the range of other developing knowledge of mental life or communication in related fields. My own research on gender and masculinity made me acutely aware that literary theory had a preferred list of theories and largely ignored most developments in the social sciences on the grounds that these presupposed dodgy concepts of mental autonomy or linguistic functionalism. Why did literary critics care so little for consistency of concepts and reasoning, and take pride in their ramshackle arguments and obvious contradictions? Even today I can find myself distracted by the mutually exclusive propositions spread over a few paragraphs in the work of one critic of poetry or another, despite the interesting ideas. Literary theory didn't see the value of shared incremental intellectual inquiry that checked and rechecked arguments against agreed data. Perhaps my renewed interest arose from the way science became more and more intrusive in our lives, as the technologies of the computer, the mobile phone and the digital camera began to alter social relations, memory and even epistemology. It was not just the writers of speculative fiction who were interested in gene expression or temporal paradoxes; poets too were writing, often with a level of understanding commensurate with a scientific education, of physics, molecular biology and the body.

These earlier interests became insistent as I worked on a long-delayed project to write about the apparently excessive ambitions of the second wave of modernist American poets who presented their books and projects as if they were as much a contribution to the creation of knowledge as the work of say Richard Feynmann, Hannah Arendt, Jacques Monod or Stephen Jay Gould. I began to think that the story of postwar American poetry was a story of both resistance to and emulation of the sciences that were encroaching on many of the areas that they felt should be natural territory for the poet. Science was increasingly the most authoritative arbiter of what was true, real, or factual, and it appeared to be reaching towards a monopoly on legitimate methods of producing knowledge even concerning the traditional fields of poetry: emotions, memory, history and social life. A poet's writing was assumed to be mere personal opinion expressed in a moody if appealing rhetoric; the scientist wrote with the authority of a vast institution that had gained access to political power and confirmed its effectiveness daily in the growing mechanization of everyday life. Every washing machine and chrome automobile fin was a reminder of who was improving the world. It would not have been surprising if poets had merely reiterated the themes of their Romantic predecessors and denounced science. Some did; there are certainly some striking polemics against the scientific world-picture to be found. Gary Snyder may write that 'science walks in beauty' in the poem 'Toward Climax' in *Turtle Island*, but his point seems to be that this beauty extends beyond and around the limited purview of the scientist.[1] His suspicions are more openly voiced in 'Mother Earth: Her Whales':

> How can the head-heavy power-hungry politic scientist
> Government   two-world   Capitalist-Imperialist
> Third World   Communist   paper-shuffling male
>   non-farmer   jet-set   bureaucrats
> Speak for the green of the leaf? Speak for the soil?[2]

Science is too complicit in the structures of state power across the globe, and too alienated from the natural world of plants and animals, to comprehend them adequately. He pictures death as a 'Liquid Metal Fast Breeder Reactor'.[3]

Other poets also raise ethical issues explicitly. In Robert Duncan's great jeremiad against Lyndon Johnson's war in Vietnam, 'Up Rising: Passages 25', the poem rails against the abdication of ethics that enables people to produce the knowledge on which the destructive weaponry depends:

> – back of the scene: the atomic stockpile, the vials of synthesized diseases eager biologists have developt over half a century dreaming of the bodies of mothers and fathers and children and hated rivals swollen with new plagues, measles grown enormous, influenzas perfected; and the gasses of despair, confusion of the senses, mania, inducing terror of the universe, coma, existential wounds, that chemists we have met at cocktail parties, passt daily and with a happy 'Good Day' on the way to classes or work, have workt to make war too terrible for men to wage – [4]

1 Gary Snyder, *Turtle Island* (New York, 1974), p. 84. 2 Ibid., p. 48. 3 Ibid., p. 67. 4 Robert Duncan, *Bending the Bow* (London, 1971), p. 82.

Duncan's apparently naïve reference to the cocktail parties (as if to say how can *my* acquaintances behave so immorally) stems from his underlying belief that this proximity is a measure of the shared cultural commonality between the poet and the scientists. The poet is as much a part of this society as the scientists who can plot the savage death of others with apparent equanimity. It is this common root that Susan Howe meditates upon in section 3, 'Taking the Forest', of *Articulation of Sound Forms in Time*. The 'corruptible first figure'[5] of the opening line has different appearances – 'Bright armies wolves warriors steer'[6] – depending on where in history one looks, whether towards religion or science. The lines – 'Vault lines divergence / Atom keystone'[7] – neatly conflate a sacred image of church roof with the reductionist particle physics of modern natural science. Howe's poem braids these and other features of the corrupted figure of humanity together until science and poetry are left with only each other for company: 'Alone in the deserts of Parchment' with the 'Theoreticians of the Modern'.[8]

These interrogations of science are only a part of the story. The leading figures within both New American Poetry and Language Poetry were not only extremely well informed about new advances in science, they also repeatedly affirmed the capacity of poetry to participate in these developments and their poetry demonstrated the depth of their convictions. These poets believed that poetry was still capable of intellectual inquiry relevant to the cultural agendas of today. Snyder draws from his reading in ecology for several poems in *Turtle Island*; science is really an age-old knowledge:

> we've always had tools and science. It was science that domesticated the goat, that developed glazed pots, that developed dyes for wool, that domesticated all the vegetables in our gardens and trees in our orchards.[9]

Duncan is more sceptical about science, and yet he too keeps it always in view. Charles Olson's *The Maximus Poems* relies extensively on archeology, and although most of us probably think of this as a humanities discipline (it is part of the School of Humanities at my university), the findings of archeological research are regularly printed in *Scientific American* and other science journals, which consider its methods of inquiry to be sufficiently scientific for inclusion. Ron Silliman, like Susan Howe, weaves scientific references into his long sequences of disjoint sentences in books such as *Tjanting*.[10]

My interest in science and poetry has a strong American bias because of my own history, but it was probably the work of two British poets, Allen Fisher and J.H. Prynne, that really crystallized the issues for me. Hearing Fisher read from

5 Susan Howe, *Articulations of Sound Forms in Time*, reprinted in *Singularities* (Hanover, NH, 1990), p. 17. **6** Ibid. **7** Ibid. **8** Ibid., p. 35. **9** Gary Snyder, *The real work: interviews and talks 1964–1979*, ed. William Scott McClean (New York, 1980), p. 146. **10** Ron Silliman, *Tjanting* (Great Barrington, ME, 1981).

Unpolished Mirrors in a reading series at Kings, I was impressed by the striking blend of dramatic monologues, London history and an echo of Olson's preoccupation with the mapping of a specific locality.[11] Here was a poetry that had learned from the Americans but didn't sound imitative. I started reading all his work, puzzled by the dense scientific language of *Defamiliarizing* that seemed almost deliberately anti-poetic, curious about the proceduralism of *Stepping*, and then startled by the depiction of South London in *Brixton Fractals*.[12] What a wonderful reimagining of London. Rarely does a poem show you a world you have glimpsed but been unable to articulate, and when this happens you feel immense gratitude and excitement. These are the opening lines of 'Banda':

> Took chances in London traffic
> where the culture breaks
> tone colours burn from exhaustion
> emphasised by wind,
> looking ahead for sudden tail lights
> a vehicle changes
> lanes into your path and birds,
> over the rail bridge, seem purple.
> A mathematician at the turn of the century
> works out invariant notions in a garden
> every so often climbs a bike,
> makes a figure of eight around
> rose beds to help concentration,
> then returns to the blackboard.
> The schemers dreamed a finite language
> where innocence became post experiential
> believing the measurable, ultra-violet from a lamp,
> isolated sunlight curvature
> made false language what can be done
> to separate
> from perception.[13]

What he calls later in this poem his 'indefinite refusal / of euphony,' and the 'quantum leap / between some lines / so wide / it hurts,' are the most visible features, but almost as striking is the repeated braiding of scientific allusions into the text as if they were as much a part of ordinary life as the traffic, the birds, or the 'sound of skateboards and rollers.'[14] The mathematician David Hilbert, whose

11 Allen Fisher, *Unpolished Mirrors* (1979–1981), reprinted in *Place* (Hastings, East Sussex, 2005). **12** Allen Fisher, *Brixton Fractals* (1985), reprinted in *Gravity* (Great Wilbraham, Cambs., 2004). **13** Ibid., p. 3. **14** Ibid., pp 6, 14.

strange activities while he worked out his logical systematization of algebra are described by Constance Reid in her biography, had a more tranquil environment in which to compose than the poet living off the walkway in Brixton that appears in many of the poems, yet both scientist and poet are imagined to be engaged in the same type of activity.[15] Reading *Brixton Fractals* I began to grasp just how important Allen Fisher considered scientific knowledge to be for the poet. His own ideas about perception and reception were constructed from his scrupulous assimilation of papers in *Nature*, the work of recent scientists, and such theories as fractals, decoherence, and chreods (he was quite willing to bring marginalized ideas into the process too), into an attempted aesthetic framework.

Allen Fisher is always willing to talk about new and exciting scientific ideas and how they might shape a poetics, so that reading his poetry I notice observations and arguments that he has discussed previously. I don't have any such access to the thinking of J.H. Prynne, but science is if anything even more salient in his poetry than in Fisher's. From the notes to 'Aristeas in Seven Years' to recent poems such as *For the Monogram*, fragments of scientific discourse adhere to poems rather in the way bus tickets and scraps of newspapers appear in a collage. Where Fisher often depicts scientific theories and facts in the process of being employed by practitioners, Prynne usually refers to some scientific concept as if the reader ought to have this as part of her or his stock of knowledge, no more esoteric than understanding the politics of the Middle East or the pollutants in food. The third section of *For the Monogram* ('7') is almost entirely written in the language of a guide to computer programming: 'At each / repeat decrement the loop to an update count.'[16] A much earlier poem, 'Acquisition of Love' from *The White Stones*, refers to the genetics of what it calls the 'neuro-chemical entail' and contrasts different models of mechanism, a lawnmower (broken and in the process of repair, this machine evokes an entire history of mower poems, as well as alluding to Martin Heidegger's discourse on tools in the early sections of *Being and Time*) with the ribonucleic acid system for maintaining and transmitting cellular forms over time.[17] Where Fisher's poetry depicts a contemporary urban landscape in which scientific knowledge is made visible as one of the determinants of the social structure, Prynne's poetry repeatedly questions the arrogated authority of science that lays claim to know not only how our bodies and environment work, but also how we should act. He makes us aware that scientific discourse increasingly plays the role once occupied by either religious or philosophical discourses.

The question that I now find myself asking of poetry that acknowledges science encompasses a much wider range than those poems that explicitly invite a dialogue with recent scientific findings. How does one investigate the impact of

15 Constance Reid, *Hilbert* (New York, 1996). **16** J.H. Prynne, *For the Monogram* (Cambridge, 1997), n.p. **17** J.H. Prynne, *Poems* (Newcastle upon Tyne, 1999), p. 111

science on poetry and discuss the ventures that poetry has carried out across domains annexed by science? I find myself wanting to narrow consideration to the modernist line in poetry because it is this poetry that starts with the assumptions that poetry can do significant cultural work, and that the poet can achieve this in part by investigating the world and the language in which we know it. How to proceed with the question is far from clear. Literary historians have become increasingly interested in tracing the history of science's impact on literature since at least the seventeenth century, and many concur that metaphors are the main currency of exchange. Others trace explicit allusions to science, images derived from the idea of black holes, or complementarity, or genetics. Just as poets once drew on neo-Platonism, or classical mythology, or Christian eschatology, they now unsurprisingly draw on the imagery employed by our new cultural masters, the scientists. Interesting as this work can be, I have found it is restrictive when studying the poetry of the past half century, partly because science has changed radically over that time.

Two features of contemporary science are especially significant. The first is the degree to which natural sciences such as physics, chemistry and biology have been transformed over the past century from a small-scale artisanal enterprise to a vast field of investment in large teams of researchers and a highly elaborated system for parcelling out shares in a co-ordinated effort and then distributing findings. The other is the way this has been accompanied by the promulgation of world views that have gained unprecedented success for such theories as the quantum mechanical picture of sub-atomic particles, or the molecular biology that treats DNA as a language (both are beginning to be revised as the character of particles gives way to superstrings, and the significance of genes switching on and off for environmental causes makes the idea of a genetic code look less convincing). This publicity for scientific paradigms often conceals the actual character of the science as it is practised in laboratories or the field, which remains poorly represented by popular accounts. Historians of the relations between science and literature have to watch that they don't rely too heavily on familiar stereotypes which when measured against scientific practice are too often a distortion of what goes on. Actual scientific knowledge at any one moment is always provisional and in transition. Listening to the media debate about global warming you might think that scientists are convinced that every new hurricane is a direct symptom of the heating of the globe but this is far from the case. The scientists who do the new research are constantly revising and even contradicting what was previously believed. Most scientists do believe that the rise in carbon dioxide levels will produce rapid global warming but they often differ radically on the details. It is widely assumed by the public that the thermohaline circulation in the north Atlantic (that includes the Gulf Stream) if halted would trigger an ice age. This was the science of the film *The Day after Tomorrow*. A few climatologists have recently argued that the cause of ice ages should be searched for in tropical air currents, and that it is not the Gulf Stream that warms Europe but the prevailing westerlies that create a maritime climate.[18] By now this theory may already be superseded. Scientific papers have relatively short citation lifetimes compared to the span of significant literary works (though much of what we misleadingly call minor poetry also has a relatively short citation lifetime and ceases to be recognized outside the most narrow circle of scholars after a few decades – look through poetry magazines or run-of-the-mill anthologies from the 1950s and you will find plenty of examples – so perhaps we should talk of the reading lifetimes of poems too).

How then can poets engage with science? The first generation of modernist poets, or high modernists, were less concerned to include scientific references than to ensure that poetry remains a participant in the cultural conversation about who we are, what the world is like, and how we should live. 'A poet is not at all surprised by science' said Louis Zukofsy,[19] and this was the attitude of such high modernist poets as William Carlos Williams who argued that the new relativistic cosmology demanded a new measure. In a lecture given in 1948 Williams demanded a new conception of poetry:

> How can we accept Einstein's theory of relativity, affecting our very conception of the heavens about us of which poets write so much, without incorporating its essential fact – the relativity of measurements – into our category of activity: the poem? Do we think we stand outside the universe? Or that the Church of England does? Relativity applies to everything, like love, if it applies to anything in the world.[20]

Hart Crane feared that the products of science might 'carve us / Wounds that we wrap with theorems sharp as hail'[21] and in a passage with resonances for Allen Fisher's 'Banda' asked, 'tell me, Walt Whitman, if infinity / Be still the same as when you walked the beach.'[22] The impressario of poetic modernism, Ezra Pound, began his textbook on poetry, *ABC of Reading* (1934), by announcing, 'We live in an age of science and abundance.'[23] He repeatedly takes his analogies for the comparative study of poems from science: 'The proper METHOD for studying poetry and good letters is the method of contemporary biologists, that is careful first-hand examination of the matter, and continual COMPARISON of one "slide" or specimen with another.'[24] In his poetry however there is little sign of scientific content; his idols are statesmen, artists, or soldiers.

18 See for example Mark Bowen, *Thin ice: unlocking the secrets of climate in the world's highest mountains* (New York, 2005), pp 307–9. **19** Louis Zukofsky, *A* (Berkeley, 1978), p. 186. **20** William Carlos Williams, *Selected essays* (New York, 1954), p. 283. **21** Hart Crane, 'Cape Hatteras', *Complete poems* ed. Brom Weber (Newcastle Upon Tyne, 1984), p. 87. **22** Ibid., p. 86. **23** Ezra Pound, *ABC of reading* (London, 1934), p. 1. **24** Ibid.

Literary critics shared the view that poetry needed to justify itself in an age of science. The literary theorist I.A. Richards, in his somewhat logical positivist account, *Science and poetry*, asks 'to what extent may science make obsolete the poetry of the past,'[25] and then observes that 'a number of men who might in other times have been poets may today be in biochemical laboratories.'[26] Notice that Richards formulates his question with the presupposition that there is an 'extent' or growth of obsolescence of earlier poetry, and raises the possibility that modern poets must modernize or risk the same fate. He might well have felt concern because although many of the modernist poets did acknowledge science in their essays, its presence in the poetry is so hard to discern. Pound makes the odd reference to Marie Curie but normally concentrates on other histories than that of science or technology. Williams' idea that his variable foot is relativistic relies on a weak analogy rather than any connection between physics and poetics (and arguably confuses relativity with relativism).

It is the American modernist poets of the 1950s who begin to work out what might be necessary for their poetry to stand alongside scientific method as a similarly valid mode of inquiry. Olson writes about poetry in a tone that conveys a belief in the importance of such work for modern society, and repeatedly draws parallels between his own poetic history and other forms of research, always urging his readers to take up their own investigations: 'I am, then, concerned as any scientist is, with penetrating the unknown.'[27] A poet of the next generation, the Language Poets, Lyn Hejinian writes about her 'romance with science's rigour'[28] and gave the title *The Language of Inquiry* to her collected essays on poetics. This was to be a poetics of inquiry, for which investigation, research, analysis, discovery, observation, reflection and other forms of intellectual curiosity were all relevant, as long as they recognized the primacy of the signifier. Her introductory essay carefully avoids prescriptive definitions of inquiry as it distinguishes *poetic* inquiry from something that sounds like the phenomenology espoused by the earlier generation, as well as the conventional picture of natural science. Watch her try to avoid making the implausible claim that poems present knowledge of the same kind as scientific papers and only partly succeed:

> Poetry comes to know that things are. But this is not knowledge in the strictest sense; it is, rather, acknowledgement – and that constitutes a sort of unknowing. To know *that* things are is not to know *what* they are, and to know *that* without *what* is to know otherness (i.e., the unknown and perhaps unknowable). Poetry undertakes acknowledgement as a preserva-

25 I. A. Richards, *Poetries and sciences: a reissue of science and poetry (1926, 1935), with commentary* (New York, 1970), p. 46. **26** Ibid., p. 52. **27** Charles Olson, 'A letter to the faculty of Black Mountain College', *Olson: the journal of the Charles Olson archives*, viii (1977) 31. **28** Lyn Hejinian, Note to 'The Person', *Mirage* 3 (1984) 24.

> tion of otherness – a notion that can be offered in a political, as well as an epistemological, context. This acknowledging is a process, not a definitive act; it is an inquiry, a thinking on. And it is a process in and of language, whose most complex, swift, and subtle forms are to be found in poetry – which is to say in poetic language (whether it occurs in passages of verse or prose). The language of poetry is a language of inquiry, not the language of a genre. It is that language in which a writer (or reader) both perceives and is conscious of the perception. Poetry, therefore, takes as its premise that language is a medium for experiencing experience. Poetic language is also a language of improvisation and intention. The intention provides the field for inquiry and improvisation is the means of inquiring.[29]

Hejinian's use of the word 'language' in the statement that 'the language of poetry is the language of inquiry' is deliberately ambiguous, meaning both metaphorically that the aims and behaviour of poetry are those of research, and literally that the actual language used in the performance of the poem is the medium of this investigation. Ron Silliman explained to the readers of the first major anthology of Language poetry, *In the American Tree*, that what united the poets was not a style or an ideology. It was common goals of inquiry:

> The nature of reality. The nature of the individual. The function of language in the constitution of either realm. The nature of meaning. The substantiality of language. The shape and value of literature itself. The function of method. The relation between writer and reader.[30]

Is 'inquiry' the link between contemporary modernist poetry such as the work of Fisher and Prynne, and science?

What does it mean to talk about inquiry in poetry? One way of thinking about this is offered by the art critic Stephen Wilson, who demonstrates that contemporary visual artists are heavily involved with the sciences. He suggests that this is more than a decorative use of scientific furniture, that the artists are playing a significant role in some cases: 'Let us define science as an accumulation of worldviews, questions, metaphors, representations, and processes that attempt to understand the nonhuman world. It is also the accumulated body of knowledge that these inquiries have generated.'[31] The question then is 'What must artists do differently than they always have done to prepare to participate in the world of research?'[32] Or, to express this in the terms of the opening question, if this is what we mean by knowledge, how can a poet write in such a way that the

29 Lyn Hejinian, *The Language of Inquiry* (Berkeley, 2000), pp 2–3. **30** Ron Silliman, *In the American Tree* (Orono, ME, 1986), p. xix. **31** Stephen Wilson, *Information arts: intersections of art, science, and technology* (Cambridge, MA, 2002), p. 39. **32** Ibid.

poetry will participate in the generation of new knowledge? Wilson's answer is that the artists need to inform themselves about science and talk to scientists. Those who have done this can produce an art that 'explores technological and scientific frontiers' and therefore can supplement scientific research by pursuing 'different inquiry pathways, conceptual frameworks, and cultural associations than those investigated by scientists and engineers.'[33] In practice, however, scientists such as the Nobel prize-winning chemist and poet Roald Hoffman, who believes that 'poetry and a lot of science – theory building, the synthesis of molecules – are […] acts of creation that are accomplished with craftsmanship, with an intensity, a concentration, a detachment, an economy of statement,' are rare.[34] Whatever supplementation is going on poets are virtually never acknowledged in this role. They have to create it for themselves, and at best call on the arguments of philosophers and historians of science who offer interpretations of science that do sometimes leave room for the activities of poetry, especially if poetry is not seen in stereotypical manner as a purely expressive, subjective deployment of metaphors and images that create non-logical connections without observational foundation.

There is ample evidence that poets think of their poems as doing more than simply reporting without addition on knowledges and findings made by others. Modernist poets do think of their poems as both reporting on their own investigations prior to the writing of the poem, and of the very writing process itself as a procedure whose outcome is not predictable and therefore constitutes a form of experiment. Should we be persuaded though? A long tradition in literary criticism rightly asks us to approach such authorial claims made in the paratexts with some scepticism. I think however it is important to point out that the text itself can make implicit claims to inquiry and that these do not fall so readily under the shadow of such scepticism. Allen Fisher's poem 'Banda' indicates to us that careful observation of the environment is to be valued: the narrator watches the traffic carefully and takes 'chances' with it, notices the colours and through verbal ambiguity suggests both that they are altered by vehicle exhausts and that his own state of fatigue affects his visual acuity; he notes that the birds appear purple though implying that he is aware that this too is a chromatic distortion; and finally mentions the unobservable radiation of the ultra-violet light. The poem also alludes to bodies of specialist knowledge and offers implicit judgements on their viability. The mathematician's eccentric behaviour contrasts with the aspiration to an absolutely reliable mathematical theory (of invariance), and those who try to develop a finite language are presented as deluded, they are both 'schemers' and dreamers. In offering these perspectives on scientific knowledge

33 Ibid., p. 3. **34** Lewis Wolpert and Alison Richards, *Passionate minds: the inner world of scientists* (Oxford, 1997), p. 23.

the poem implies that it too has the capacity to put its own reasoning and research alongside the discourse that engages with these theories.

Poets find science interesting. Its objects of attention also interest poets. They too observe changes in the wilderness or trace patterns of human behaviour, but they do this work in and through language, though that may say less than we imagine because we need to make a distinction between language as a practice and the various and changing concepts of it, ideas that sometimes have been treated by literary theory not as the models they are but as adequate substitutes for the actuality, the entire panoply of activities that involve speech, communication, writing, art and records. Poets ask questions and try to find out the answers by methods that may be raw versions of the methods used by scientists and may not always share them in a similar manner to the scientific networks, nor are they confined to working within an existing paradigm to anything like the same extent. Individual scientists each work on a tiny part of a big problem and have extremely refined means of sharing their results with each other so that a picture that commands reasonably widespread consensus can emerge. This co-ordination requires writing that is not only unambiguous but generates agreed inferences and is calibrated so that the degree of certainty or speculation can be read off from the rhetoric of the scientific paper. Modernist poets do however co-ordinate their efforts more than we have usually recognized, and although they enjoy ambiguity they do so intentionally, using the polyvalency of the lexicon and the many semantic resonances that recede away from a word or phrase via such different pathways as etymology to phonemic resemblances and sound patterns.

If we assume that poetry can itself be a mode of inquiry pursuing alternative pathways of research that are nevertheless equal in value to science, how do we produce an unequivocal statement of the achievement of a poem? Does Allen Fisher discover that there are cultural discontinuities evident in the life-world of the British capital, for instance? In practice we either have to point to innovative use of language (in Fisher's case perhaps the distinctive way in which sentences from quite different domains are juxtaposed, the painful quantum leap) and experiments in form, or to provisional statements in the poem that indicate observations, ideas, or self-analyses that add to a wider debate on the specific theme in question. Most commentators opt for the first route because poetry appears to offer no statements with the degree of assurance that would in any way be commensurate with a scientific discourse. Poems appear to make no claims to knowledge, offer no commitments or judgements, and don't speak with any of the authority that is invested in a scientist, a scholar or anyone speaking in an official capacity. Poems are full of ambiguity; unfinished and therefore unasserted propositions are left suspended in evanescence. Postmodern criticism presents this as play and ridicules any claim to truth in such aesthetic discourse.

Before I try to show that there might be tangible results of poetic inquiry even in a paratactic and rhetorically ambiguous poem, I want to return to the point that science is itself much more fluid and provisional than its media image allows, and does not as it were explain everything. Love, numbers and bodies cannot all be treated as the direct causal outcome of tiny particles moving according to the laws of physics. There is lots of room for inquiry that is not scientific. As the philosopher of science John Dupré says, 'there are surely paths to knowledge very different from those currently sanctioned by the leading scientific academies.'[35] It is a mistake to think that the world revealed by science is the one in which we actually live rather than a cleverly constructed abstraction from it. The philosopher Wilfrid Sellars contrasts what he calls 'the scientific image of man', with the everyday image of the world, 'the manifest image' and argues that this and the distinctively scientific one are not to be thought of either as mutually exclusive alternatives, nor as is tempting, to treat them as complementary parts of a duality, science here and everyday thought there. He argues that:

> The conceptual framework of persons is not something that needs to be *reconciled with* the scientific image, but rather something that needs to be *joined* to it. Thus to complete the scientific image we need to enrich it *not* with more ways of saying what is the case, but with the language of community and individual circumstances in which we intend to do them in scientific terms, we *directly* relate the world as conceived by scientific theory to our purposes, and make it *our* world and no longer an alien appendage to the world in which we do our living. We can, of course, as matters now stand, realize this direct incorporation of the scientific image into our way of life only in imagination.[36]

The American modernist poets have been doing this, incorporating the scientific image and in the process often correcting or adjusting it.

Another cause of our difficulty in grasping what poetic inquiry might achieve is that literary criticism and literary theory have generally treated literary writing as not subject to the same demands for truth and accuracy as other modes of writing. A literary critic's essay is supposed to meet certain criteria: it should provide evidence for its assertions, evidence that is checkable and accurately reported, it should offer rational argument whose steps can be reconstructed, and it should cite its data and sources accurately. Such criteria are so accepted in the fields of history or sociology say, that they hardly need mentioning. Even in philosophy, where speculative argument is prized above almost everything else, codes of good

35 John Dupré, *The disorder of things: metaphysical foundations of the disunity of science* (Cambridge, MA, 1993), p. 10. **36** Wilfrid Sellars, 'Philosophy and the scientific image of man', in *Science, perception and reality* (Atascadero, CA, 1991), p. 40.

practice ensure that the sources of an idea are credited and that they are not misrepresented, and that clarity is wrested from the ever present risks of ambiguity and indeterminacy. The field of literary studies has long accepted almost without question that a literary work is free to exploit linguistic indeterminacy and play with ambiguity and fictiveness because the work does not and could not legitimately make claims about the truth or accuracy of its representations. Jean-Paul Sartre's argument that most literary fictions start out as committed works deeply embroiled in the controversies of their own day and only later achieve the seeming harmonious detachment that we call literary, remains a rare exception to this rule, and even he could see no such realism in poetry. In practice literary analysis often does take for granted that literary writers are striving to investigate the world they depict and to do so with accuracy and insightful discovery even if they do so through the obliquities of abstraction, metafiction, allegory, symbolism, or fantasy. Without such an implicit assumption most historicist criticism would be irrelevant. It remains difficult however to see how we might begin to acknowledge the significance of truth and accuracy in literary writing, especially poetry, given the fictiveness and semantic density arising from the texture of sonic, visual, contextual and etymological cues at work.

One crucial implication of Sellars' argument is that it would be inappropriate to assume that if poets produce knowledge (that may belong to a subset of the manifest image perhaps, a more specialized version than the everyday common sense one), this knowledge cannot be evaluated by checking how well it corresponds with the state of affairs described by the latest science. The poets' knowledge is better thought of as depending upon the beliefs on which ordinary knowledge is continually reassessed. These are the reasonings that comprise the current state of knowledge of our world in its entirety (and not just the scientific image of it). One consequence of this is that we might do better to think of what the poets produce in terms of inference rather than representation.

Analyzing Fisher's poem 'Banda' I resorted several times to the innocuously familiar concepts of implication and inference. I have in effect been saying that it is reasonable to infer that Fisher's poem not only values observation and scientific reasoning but that the text itself is aligning itself with those inquiries, offering its own specific contribution to them. The concept of inference is used deliberately to allude to recent developments in the philosophy of language that I find helpful for thinking about the achievements of poetic inquiry. Much of the confusion about truth and knowledge in poetry arises because we assume that the only mode in which a poem could demonstrate that its inquiries not only were capable of but actually had produced knowledge and certainty of the kind that we value science for, would be through representation and its simplest form reference. A poem claims something and we can, at least in theory, check it against the best available knowledge. The difficulty we then have is two-fold at least. Poetic textuality is almost always complexly ambiguous and it is rarely possible to attribute a singular proposition to a line or sentence in a poem, let alone to the entire poem, made up as it is likely to be of sentences and phrases that are not seamlessly hypotactic. If we are looking for an unequivocal representation of a state of affairs we won't find it. Instead we have to infer what usually turn out to be several alternative possibilities and these in turn, as Paul de Man brilliantly demonstrated in his essays, can sometimes even be mutually cancelling. If we do find a statement that is reasonably monovalent and can be taken as a checkable statement about states of affairs, the referent itself turns out to be oddly elusive. Is Fisher's mathematician a symbolic figure, or the historical person the facts of whose life are recorded in the biography, or is he possibly some other even fictional character (there is nothing in the poem to fix this as Hilbert though the notes at the end of the book would allow us to assume this because they give a reference to his biography)? These are only a few of the possibilities. Reference depends radically on context, and as we frequently are reminded, even with a rich contextual framework, the target of a reference can be mistaken or become ambiguous. A poem like this provides lots of latitude for possible inferences.

Recent philosophical arguments about language by Robert Brandom and others may help us to reach a better picture of what the relation of poetry to science might be. Brandom has developed the ideas of Sellars and others who argue that it is more fruitful to think of meaning or semantics in terms of inference rather than representation. We should be asking why we think that a particular statement appears referential, or how a word can represent something. Brandom's philosophy is complex and even a brief account is beyond this essay. Its core idea does however suggest that one of the main reasons why we find it so hard to see anything in common between scientific discourse and poetry is that we may be assuming that the truths of science depend on their correspondence to facts or the state of the world. If instead we were to think of knowledge as a form of commitment to a belief to which the believer is entitled and which then entails a similar commitment from the listener or reader, we might see how poems too could in certain circumstances create knowledge. Brandom insists that 'taking a claim or belief to be true is not attributing an especially interesting and mysterious quality to it; it is doing something else entirely. It is *endorsing* the claim oneself.'[37] Everyday use of language relies on the fundamental significance of inference rather than representation because to speak or write is to make judgements whose commitments and consequences are then negotiated with listeners and readers. A concept is at heart a commitment to a potential network of inferences from such a starting point. This theory of language enables us to speculate that what is happening in a poem like Fisher's is that the inferential patterns are altered from their

37 Robert Brandom, *Articulating reasons: an introduction to inferentialism* (Cambridge, MA, 2000), p. 119.

more normative uses, but they have not been deleted. There is plenty of inference at work that invites reasoning and possible response, whether affirmative or resistant, a response that would lead to further explicitness and new inference networks. As Brandom writes:

> The significance of a speech act is how it changes what commitments and entitlements one attributes and acknowledges [...] [D]iscursive commitments (to begin with, doxastic ones) are distinguished by their specifically inferential articulation: what counts as evidence for them, what else they commit us to, what other commitments they are incompatible with in the sense of precluding entitlement to.[38]

In his view it is a mistake to think that the meaning of a sentence depends on what it refers to or represents; its meaning arises from the inferences that can be drawn from it, and the kinds of justification it gives rise to.

Brandom's philosophy of semantics may turn out to be in need of radical revision, or require other philosophical entailments that we don't want to make. Its existence should be enough however to remind us that there is still much about the way language and argument work that is not well understood and remains in need of investigation, and therefore the claims by poets that their poems may operate within fields of inquiry that intersect with science may not be merely exaggerated claims to importance; they may yet find articulation within emerging theories of language and literature. I don't want to leave the impression that Brandom's philosophy is another of those magical solutions that literary theory used to be so good at producing out of the convoluted translations of enigmatic French or German prose. It suffices here to introduce a modicum of doubt that the apparent failure of poetry to provide representations of the world that are consonant with those of science is sufficient to disqualify all attempts to treat poetry as a parallel endeavour to modern science. If knowledge is based on a web of inference, then it might be possible to analyse in detail how recent poetry can claim to be a language of inquiry, and to support detailed readings of individual poems.

Poets have already tried to show us what this might be like. In conclusion, I shall return to Robert Duncan and his best-known poem, 'A Poem Beginning With a Line by Pindar.' Pindar and 'the old poets', and Walt Whitman himself, are all invoked as guides to survival in a warlike, materialist age, and Duncan repeatedly asks a similar question to Crane's question about the permanence of 'infinity', this time in the form of a question about the myth of Eros and Psyche: 'What if they grow old?'[39] Even the poet who, if he is an heir to Whitman, must

38 Ibid., p. 81. **39** Robert Duncan, *The Opening of the Field* (New York, 1960), p. 63.

also 'contain multitudes', is affected: 'I too / that am a nation sustain the damage.'[40] His subtle use of the myth of Eros and Psyche is a way of insisting that modern science has not set aside all the knowledge that belonged to past cultures as triumphalist theories of the growth of science would propose. The objective methods of science do not have exclusive epistemological validity. By concentrating on the issues of belief and representation raised by the discourse of the 'seraphic', Duncan avoids thinking of science as a given reality whose only language is technology. At one point he calls Psyche 'Scientia' (knowledge) and by doing so allows the modern word 'science' to be heard, so that the entire poem can be read as a meditation on what ought to constitute a modern science:

Scientia

holding the lamp, driven by doubt;
Eros naked in foreknowledge
smiling in his sleep;   and the light
spilld, burning his shoulder – the outrage
        that conquers legend –
passion, dismay, longing, search
        flooding up where
the Beloved is lost.   Psyche travels
life after life, my life, station
        after station,
to be tried

        without break, without
news, knowing only – but what did she know?[41]

What, Duncan is asking, does modern *scientia* know and what are the dangers and consequences of such attempts to know? What knowledge does poetry create? All this takes place not in a timeless realm either, but in a modern America of fallible government:

Hoover, Coolidge, Harding, Wilson
hear the factories of human misery turning out commodities.
[. . . . .]
Harding, Wilson, Taft, Roosevelt,
idiots fumbling at the bride's door,
hear the cries of men in meaningless debt and war.
Where among these did the spirit reside
that restores the land to productive order?[42]

40 Ibid., p. 64. **41** Ibid., p. 66. **42** Ibid., p. 64.

And I might add, John F. Kennedy and Lyndon Baines Johnson, taking America to war with Vietnam and nearly to war with Russia, when I first became interested in science and literature. Duncan rightly insists that science needs literature or legend if it is to avoid reducing human life to a meaningless material struggle within the 'factories of human misery'. Poetry has a responsibility to the *scientia* at its heart, and if it attends to this can also find a way to be scientific in ways that do collaborate imaginatively with the sciences of today.

Part II

Science and poetry: not so different?

IGGY McGOVERN

When the cube and things together
Are equal to some discreet number,
Find two other numbers differing in this one.
Then you will keep this as a habit
That their product should always be equal
Exactly to the cube of a third of the things.
The remainder then as a general rule
Of their cube roots subtracted
Will be equal to your principal thing
In the second of these acts,
When the cube remains alone,
You will observe these other agreements:
You will at once divide the number into two parts
So that the one times the other produces clearly
The cube of the third of the things exactly.
Then of these two parts, as a habitual rule,
You will take the cube roots added together,
And this sum will be your thought.
The third of these calculations of ours
Is solved with the second if you take good care,
As in their nature they are almost matched.
These things I found, and not with sluggish steps,
In the year one thousand five hundred, four and thirty.
With foundations strong and sturdy
In the city girdled by the sea.[1]

This strange poem was written in 'the year one thousand five hundred, four and thirty', as the speaker says, by the mathematician Niccolo Fontana (*c.*1499–1557),

1 The text of Tartaglia's poem with detailed commentary on his life and work is available online at http://www-history.mcs.st-andrews.ac.uk/HistTopics/Tartaglia_v_Cardan.html, accessed 16 July 2006.

better known as 'Tartaglia' ('the stammerer'). Perhaps it sounds more 'poetic' in its original Italian; in translation, despite the final flourish, it is not hard to guess what it represents: the solution of a particular type of cubic equation called 'unknowns and cubes equal to numbers' or, in modern algebraic notation, $x^3 + ax = b$. Already, though, I sense the uneasiness of the reader: why would anyone write a poem about solving an equation? The answer is simple enough. 'Tartaglia' wanted to keep his discovery secret. Certainly, the idea of using a poem as a cipher should not be unfamiliar to the reader of poetry. Indeed, one of Ireland's longest established literary magazines is called *Cyphers*.

I have two purposes, then, in beginning with this poem. First, I want to raise an early flag for 'science' in a collection that *may* appear to lean more towards the other side of the house, as it were. Secondly, and more importantly, this essay tries to address the question raised in the title in a more general way. Here are two disciplines (for want of a better word), 'science' and 'poetry', two disciplines that at first sight seem to have little in common. Physically, in Trinity College Dublin at any rate, they occupy opposite ends of the campus (although the Oscar Wilde Centre for Creative Writing maintains a Treasure Island-like stockade in the Science concourse of TCD). American-styled 'Physics for Poets' courses apart, students with interests in both subjects must often make the hard choice between them. All of this suggests that they are poles apart, but I will try to show in this essay that they are not in fact so very different. My method probably owes more to 'poetry' than to 'science', in that it involves a fair degree of circling around the topic, rarely reaching definitive conclusions, but hopefully I will draw attention to other possibilities not often acknowledged in the debate.

* * *

One positive sign that needs to be acknowledged at the outset is the recent appearance of a number of science-poetry anthologies. But consider this quotation from Maurice Riordan and Jon Turney, the editors of one such anthology, *A Quark for Mister Mark*: 'The notion that there is any deep relation between science and art is, in any case, belied by the fact that there are vanishingly few distinguished practitioners of both.'[2] As a 'practitioner of both' myself, I hope this dismissal is not the only reason that I find this anthology disappointing. *A Quark for Mister Mark* has the subtitle *101 poems about science*. Certainly there are 101 poems in it, many of them excellent, but I am at a loss to see how any more than a few of them could be described as being 'about' science. A much more satisfying anthology is the earlier *Poems of science*, edited by John Heath-Stubbs and Phillips

2 Maurice Riordan and Jon Turney (eds), *A Quark for Mister Mark: 101 poems about science* (London, 2000), p. xiii.

Salman.[3] That it is better in its focus on science is due in part to its much longer reach, from the thirteenth century to the 1980s. One difficulty for the Riordan and Turney anthology is that it confines itself to a narrow time window, beginning with the publication of *Poems of science*. But it also suffers from the self-imposed limitation of excluding poetry in translation. Thus, there is only a token poem by Miroslav Holub, one of the famously few distinguished contributors to both science and poetry. I suspect that while the editors may indeed 'know a good poem' they know rather less about science.

I know of two further anthologies which are more scientifically grounded. *Verse and universe*, edited by Kurt Brown, is a substantial selection of modern American poetry with considerable science content.[4] The poems are usefully grouped under headings such as 'space', 'time', 'matter', and so on. David Morley and Andy Brown, the editors of another anthology entitled *Of science*, are poets with scientific backgrounds, as are all the contributing poets.[5] Moreover, the poems are intentionally unattributed so that 'the project should stand or fall on the selection of the language of science rather than on the reputation of its writers.'[6] An important companion to these anthologies for anyone interested in the discussion is Holub's collection of essays, *The Dimension of the Present Moment*.

In his essay 'Poetry and Science', Holub playfully questions the issue of the relationship between science and poetry:

> Why, then, should it make so much difference, being the poet and being the scientist, when 95 per cent of our time we are really secretaries, telephonists, passers-by, carpenters, plumbers, privileged and underprivileged citizens, waiting patrons, household maids, clerks, commuters, offenders, listeners, drivers, runners, patients, losers, subjects and shadows?[7]

But what about the other 5 per cent of the time? Ezra Pound's instruction to poets to 'make it new' will not be unfamiliar to scientists. But there are those who would scorn this connection. Consider, for example, the following claim by the physicist and Nobel laureate Paul Dirac:

> How can you do physics and poetry at the same time? The aim of science is to make difficult things understandable in a simpler way: the aim of poetry is to state simple things in an incomprehensible way. The two are incompatible.[8]

3 John Heath-Stubbs and Phillips Salman (eds), *Poems of science* (London, 1984). 4 Kurt Brown (ed.), *Verse and universe* (Minneapolis, 1998). 5 David Morley and Andy Brown (eds), *Of science* (Tonbridge, 2001). 6 Ibid., p. 5. 7 Holub, *The Dimension of the Present Moment* (London, 1990), p. 145. 8 Helge Kragh, *Dirac: a scientific biography* (Cambridge, 1990), p. 258.

It is hard to believe that Dirac could have been as rude as this. After all, he is often regarded as the 'arty' physicist: he is the originator of the concept of the 'beautiful' as opposed to the 'right' equation, the one that merely fits the experiments. But even if he was being ironic, the question of 'physics and poetry at the same time' still lingers …

A balance might be easier to achieve if there was a proper word for it – and I believe there is. The term 'sunlighter' is now used to describe someone with two careers, but openly so, unlike the 'moonlighter', who does his 'other' work under the cover of night. It is not too big a leap from 'sunlighter' to 'rainbows'. For if anything symbolizes the science-poetry divide, in my view, it is the rainbow: 'My heart leaps up when I behold / A rainbow in the sky', William Wordsworth wrote.[9] The first-generation Romantic Wordsworth was not so down on science, provided, of course, it knew its place. The second-generation romantic John Keats was more explicit, however, complaining that Isaac Newton had destroyed the poetry of the rainbow by reducing it to a prism.[10] 'Stop messing with our rainbows' became the rallying cry! But did these 'rainbow warriors' Wordsworth and Keats never notice the secondary rainbow? If so, did they not see that the order of colour across the arc was reversed in this secondary rainbow? Or, looking more closely, did they not observe that the colour-structure also varied *along* the arc? For the rainbow is just the entrée to the wonderful world of spectroscopy. But this is beginning to sound like a science talk, so I'll leave it to the popular science writer Richard Dawkins, author of *Unweaving the rainbow*, who argues that: 'Keats could hardly have been more wrong […] Science is, or ought to be, the inspiration for great poetry.'[11]

So, how successful has this inspiration been? According to the editors of *A Quark for Mister Mark*, 'Poets are good it seems on heavenly bodies and lower life forms.'[12] The point seems to be supported by Paul Muldoon, in his poem 'Lunch with Pancho Villa' where he imagines the Mexican poet 'rambling on, no doubt, / About pigs and trees, stars and horses.'[13] While we're on the subject of 'lower life forms', though, consider the following fragment from the seventh-century BC poet, Archilochus: 'The fox knows many things, but the hedgehog knows one big thing.' But what about the poetry of 'the stars'? Today's astronomers are themselves the 'big stars' of physics; everything is big about them – their distances, their numbers of stars, galaxies; even their bang is big! So, I feel a guilty pleasure in seeing them get their come-uppance from a poet who might be regarded, after Wordsworth and Keats, as a third-generation Romantic: Walt Whitman. In his

9 William Wordsworth, 'The Rainbow', in Duncan Wu (ed.), *Romanticism: an anthology* (Oxford, 1994), p. 367. **10** See John Keats, 'Lamia', in Wu, *Romanticism*, pp 1064–79. **11** Richard Dawkins, cited online at http://www.simonyi.ox.ac.uk/dawkins/WorldOfDawkins-archive/Dawkins/Work/Books/unweaving.shtml, accessed 2 August 2006. **12** Riordan and Turney, p. xiii. **13** Paul Muldoon, *Mules* (London, 1977), p. 13.

poem 'When I heard the learn'd astronomer' Whitman ponders the astronomer's impressive display of proofs, figures, charts and diagrams, before delivering his final dismissal:

> When I sitting heard the astronomer where he lectured with
> much applause in the lecture-room,
> How soon unaccountable I became tired and sick,
> Till rising and gliding out I wander'd off by myself,
> In the mystical moist night-air, and from time to time,
> Look'd up in perfect silence at the stars.[14]

The poet and the physicist in this poem seem worlds apart, but appearances, in poetry as well as in physics, can be deceiving …

* * *

In his essay 'Poetry and Science' Miroslav Holub begins by observing that what really distinguishes poetry from science is what seems to unite them: words. 'Writers use the same tools (i.e. words) as scientists' he argues: 'They perform on the same stage, but move in the opposite direction. The sciences and poetries do not share words, they polarize them.' But is it not just a matter of unshared *meanings* of words? The 'sciences', Holub responds, 'are based on a *single* logical meaning of the sentence or of the word […] On the contrary, poetry tries for as *many* possible meanings and interactions between words and thoughts as it can.'[15] Echoing Dirac, the reductionist and the expansionist can have no commonality even when the individual words mean the same to both. And now I am thinking again of our friend Archilochus: who is the fox and who is the hedgehog here? In particular, who is who when we consider the possible consequences of either poetry or science?

Holub quotes the critic I.A. Richards, who suggested that '[t]he poetic approach evidently limits the framework of possible consequences […] For the scientific approach this framework is unlimited.' It is the pared-down scientific statement that has the greater range of possible consequences – an instance of the uncertainty principle, perhaps? – and therefore to my mind the fox is the poet, while the scientist is the hedgehog. But then Holub turns to discuss the commonalities of science and poetry. He identifies three aspects, which I call 'deciding', 'doing' and 'finding'. The first of these aspects is 'deciding' what experiment to do or what poem to write, a process which 'starts with a heavy burden of accumu-

14 Walt Whitman, *Leaves of Grass & other writings*, ed. Michael Moon (New York and London, 2002), p. 227. 15 Holub, *The Dimension of the Present Moment*, pp 131, 133.

lated literature,' but is driven by 'an acute enthusiasm.' All serious scientists and poets will recognize this process, and what Holub sees as the 'acute enthusiasm' of research in particular. The next phase of the project is the 'doing' of the experiment, or the writing (making) of the poem. This, Holub suggests, involves 'discipline' and 'tolerance for pitfalls'. It is also 'a lonely stubborn and defensive endeavour' which presents 'the basic risk of losing'. Again, it can be said that this process applies as well to the writing of poetry as it does to scientific research. Finally, Holub gives us the idea of 'finding' or what might be called the 'eureka' moment. In his words this involves '[t]he experience of the little discovery' and it 'is the same when looking into the microscope and when looking at the nascent organism of the poem. It is one of the few real joys of my life.'[16] (Or of anyone's life!) And there is a further point of contact in the ultimate value of such exercises. Although, as stated earlier, 'poetry and science move in [...] opposite directions', at the same time 'they do not aim [...] for opposite ends.' Their common end is 'saving', which is the relieving of oppression – either of a 95 per cent quotidian existence or of totalitarianism, whether of professional hierarchy or of the state. Holub's personal witness to the latter informs one of his most memorable poems, 'Žito the Magician', in which the eponymous figure must amuse his royal employer by changing water into wine, frogs into footmen, beetles into bailiffs, as well as thinking up such seeming absurdities as black stars or dry water. But then 'along comes a student and asks: Think up sine alpha / greater than one':

> And Žito grows pale and sad: Terribly sorry. Sine is
> between plus one and minus one. Nothing you can do about that.
> And he leaves the great royal empire, quietly weaves his way
> Through the throng of courtiers, to his home
>
> in a nutshell.[17]

II

In this part of the essay, I want to be more local, less theoretical, almost experimental. I will start by revisiting the idea of a 'poem about science'. Is there such a thing? Tartaglia's poem notwithstanding, can you have a poem which actually contains some science? According to Heath-Stubbs and Salman in their introduction to *Poems of science*, if there is such a thing it is a rarity. They argue that:

> The poem of science that directly incorporates quantitative or formulaic scientific material has to remain an aloof sub-genre within the larger genre

16 Ibid., pp 134, 138, 139, 141, 143. **17** Holub, *Poems Before and After*, trans. Ian and Jarmila Milner, Ewald Osers, and George Theiner (Newcastle upon Tyne, 1990), p. 69.

> we represent here until and unless scientific material becomes more like 'common knowledge'. That is, the sub-genre will exist only where there are writers and readers who have scientific knowledge.[18]

In this regard the later anthology *Of science* is significant because, as the editors explain, '[t]he writers of this selection learned their attitudes of science "in laboratories", in the fields of freshwater ecology, mathematics, marine biology, neural physiology, ethnology, computing, phenomenology and biochemistry.'[19] Nor is the list exhaustive. The late Rebecca Elson (1960–99) was an astronomer and the opening poem of her posthumous first collection *A Responsibility to Awe* is something of an antidote to Whitman's poem quoted earlier. She herself is one of the 'nomads, / Merchants, circus people' she describes in her poem 'We Astronomers':

> Sometimes, I confess,
> Starlight seems too sharp,
>
> And like the moon
> I bend my face to the ground,
> To the small patch where each foot falls,
>
> Before it falls,
> And I forget to ask questions,
> And only count things.[20]

But in case this looks like a thoroughly modern phenomenon, let me introduce a much older (and Irish) voice: William Rowan Hamilton.

Hamilton is unquestionably the most important Irish scientist ever. The year 2005 saw the bicentenary of his birth and scientists in Trinity College Dublin were put on notice to mention Hamilton's name on every conceivable occasion! In this instance compliance is easy, for Hamilton also wrote poetry. Although he was Royal Astronomer and Director of Dunsink Observatory in Dublin, his real interests were more mathematical and poetical. For example, in 1848 he visited Parsonstown in Birr, County Offaly, the home of Lord Rosse and the site of the then largest telescope in the world; but while the others in the party were engaged in discussions astronomical, Hamilton had, Whitman-like, 'wander'd off' to compose a sonnet in praise of the telescope's builder:

> I stood expecting, in the Gallery,
> On which shine down the Heaven's unnumbered eyes

18 Heath-Stubbs and Salman, p. 37. **19** Morley and Brown, p. 5. **20** Rebecca Elson, *A Responsibility to Awe* (Manchester, 2001), p. 9.

Poised in mid air by art and labour wise,
When with mind's toil mechanic skill did vie,
And wealth free poured, to build that structure high,
Castle of Science, where a Rosse might raise
(His enterprise achieved of many days)
To clustering worlds aloft the Tube's bright Eye.
Pursuing still its old Homeric march,
Northward beneath the Pole slow wheeled the Bear;
Rose overhead the great Galactic Arch
Eastward and Pleiades with their tangled hair
Gleamed to the west, far seen, the Lake below;
And through the trees was heard the River's flow.[21]

Poetry of its time, certainly, and is there not the faintest echo in that final couplet of Samuel Taylor Coleridge's 'Kubla Khan'?[22] Hamilton is important not just for the writing of poetry, though, but also for his interaction with the poets of the day, including Coleridge, Wordsworth and 'Speranza', the mother of Oscar Wilde. Moreover, he had strong views on the synergy of science and poetry. 'With all the real differences between Poetry and Science,' he once said, 'there exists, notwithstanding, a strong resemblance between them; in the power which both possess to lift the mind above the stir of the earth, and win it from low-thoughted care.'[23] Hamilton stuck mainly to the day job, however, and while he continued to write poetry, his chief focus was on science. Had it been otherwise, might he have missed his 'eureka' moment, the famous discovery of 'quaternions' on the Royal Canal when he 'felt the galvanic circuit of thought close' and he could not 'resist the impulse [...] to cut the formula on a stone of Brougham Bridge'? But Hamilton remained true to his belief that science and poetry were inseparable, remarking that the same quaternions had four parents: geometry, algebra, metaphysics – and poetry.[24]

In the following poem, therefore, I have contrived to make Hamilton re-tell the story in a form known as *bouts-rimes*. Making *bout-rimes* was a popular nineteenth-century parlour game of verse-making based on the end-words of a well-

21 For a more detailed discussion of this and other poems by Hamilton see the following website: http://www.rte.ie/radio1/story/1065121.html. **22** Like 'Tartaglia', Coleridge was also interested in the relationship between mathematics and poetry. See 'A Mathematical Problem' in Ernest Hartley Coleridge (ed.), *The poems of Samuel Taylor Coleridge* (London, New York, Toronto, Melbourne, and Bombay, 1912), pp 21–4. An early draft of this poem, dated 17 March 1791, is given the following subtitle: 'Prospectus and specimen of a translation of Euclid in a series of Pindaric odes, communicated in a letter of the author to his brother Rev. G. Coleridge.' **23** William Rowan Hamilton, 'Introductory lecture on astronomy', delivered in Trinity College, Dublin, 8 November 1832. Reprinted in the *Dublin University Review and Quarterly Magazine* 1 (January 1833), 72–85. Available online at http://www.maths.tcd.ie/pub/HistMath/People/Hamilton/Lectures/AstIntro.html. **24** For a detailed discussion about Hamilton's discovery of quaternions, see the following website: http://www.maths.tcd.ie/pub/HistMath/People/Hamilton/Quaternions.html.

known poem. In this case I have used a poem by another canal-man, the twentieth-century Irish poet Patrick Kavanagh, to make a poem about Hamilton. It's called 'Sir William Rowan Hamilton's Bouts-Rimes':

> Commemorate me, too, where there is water.
> Though I might favour 'current' over your 'stilly
> greeny', I salute you, Paddy, a brother
> poet – O to have written as beautifully!
> (Speranza would have loved 'Niagarously roars'!)
> Poor Wordsworth used to sink into cold silence
> or turn the conversation towards prose
> when I dared 'versify': 'Parnassian islands'
> would garner his profuse apologies.
> I carved *my* finest poem on Brougham Bridge's
> stonework, seized by some wild empathy
> to charm Medusa-like mythologies
> from the symbols (i, j, k), hero-courageous
> graffiti wasted on the passer-by.[25]

It is not far from the play of *bout-rimes* to the work of the *Ouvroir de littérature potentielle* or 'Workshop of Potential Literature'. This, or Oulipo, is the name of a group of mainly French writers, mathematicians and academics who have combined mathematics and literature to generate new literary structures.[26] Their aim, or indeed *bout*, would surely have met with Hamilton's approval. For example, *Cent mille milliards de poèmes* (translated as *One Hundred Trillion Sonnets*) by Raymond Queneau is a ten-page work with a sonnet on each page, but the pages are cut crosswise so that each line of each sonnet can be turned separately. Any of the ten first lines can be combined with any of the ten second lines to give one hundred (or 10^2) different pairs of opening lines. Including all fourteen lines of each sonnet, there are 10^{14} or one hundred trillion possible combinations! Queneau claimed that each sonnet made sense but, as the mathematician John Allen Paulos has commented, this claim will never be verified since there is more text in this 'collection' than in all the rest of the world's literature.[27]

This techno-poetry reminds me that even within science there may still be found traces of an old idea, that of 'science-good'/'technology-bad'. How comforting to have this prejudice confirmed by the poets! In their introduction to *A Quark for Mister Mark* Riordan and Turney say that, 'although [their] definition of science is inclusive, [they] found rather little to read about technology'.[28]

25 This poem is also published in Iggy McGovern, *The King of Suburbia* (Dublin, 2005), p. 44. **26** The following website contains a useful introduction to the work of the Oulipo group: www.nous.org.uk/oulipo.html. **27** See www.math.temple.edu/~paulos/ for John Allen Paulos's discussion of Queneau's work. **28** Riordan and

Certainly they missed one excellent example of technology-inspired poetry from Robert Crawford's 1990 collection *A Scottish Assembly*. In his poem 'Scotland' Crawford equates his homeland with the technology of the microchip, reflecting Scotland's status as the Silicon Valley of these islands:

> Semiconductor country, land crammed with intimate expanses,
> Your cities are superlattices, heterojunctive
> Graphed from the air, your cropmarked farmlands
> Are epitaxies of tweed…

I use the poem in my lectures on Epitaxial Physics and even the most hard-bitten technologist cannot be unmoved by its concluding lines:

> Among circuitboard crowsteps
> *To be miniaturised is not small-minded.*
> To love you needs more details than the Book of Kells –
> Your harbours, your photography, your democratic intellect
> Still boundless, chip of a nation.[29]

Crawford is Professor of Modern Scottish Literature at the University of St Andrews. One of his innovations has been the pairing of poets and scientists, nine pairs in all. The one pairing that I want to focus on concerns the Irish poet, Paul Muldoon. Muldoon is Professor of Creative Writing at Princeton University and he was paired with the Professor of Chemistry at the same university, who rejoices in the rather poetic name of Warren S. Warren. (It is not without significance that the pair had not previously met.) Muldoon visited Warren's research laboratory and subsequently wrote a poem called 'Once I Looked into Your Eyes'. Warren's research area is magnetic resonance and his response to the poem was to set up an experiment based on magnetic resonance imaging. He took scans of the brain of a volunteer reading a Muldoon poem, interspersed with university administrative regulations. These showed significantly more enhanced prefrontal cortex activity when the reader hit the poetry. I can't help feeling that the better experiment would have been to scan the poet's brain during the composition of the poem, which ends:

> I'm somewhat gratified to find
> That by laser-enhanced
> Magnetic resonance,
> If nothing else, I may read your mind[30]

Turney, pp xiii–xiv. **29** The poem is included in Michael Hulse, David Kennedy, and David Morley (eds), *The new poetry* (Newcastle upon Tyne, 1993), p. 278. **30** The poem is available online at www.princeton.

But there is also good news for Schools of English when Warren comments that brain sensors will not replace literary critics for some time!

Speaking of literary criticism, I can't resist including the following from Kurt Vonnegut, in his recent memoir *A Man Without a Country: A Memoir of Life in George W. Bush's America* (2006):

> I know that customarily English Departments in universities, without knowing what they are doing, teach dread of the Engineering department, the Physics department, and the Chemistry department. And this fear, I think, is carried over into criticism. Most of our critics are products of English departments and are very suspicious of anyone who takes an interest in technology. So, anyway, I was a Chemistry major, but I'm always winding up as a teacher in English departments, so I've brought scientific thinking to literature. There's been very little gratitude for this.[31]

I imagine that somewhere behind that piece lurks the old saw that it is the job of the arts to civilize the sciences. Thankfully, it is an idea that has little currency, and I want to tell its remaining adherents not to worry, that the scientists are more than capable of doing that job themselves, as the following example from Holub illustrates. In his poem, 'Brief reflection on cats growing in trees', he relates how moles apply the scientific methods of research to determine what is above ground: the mole sent up to investigate catches sight of a tree with a bird on it, leading to the theory that up above birds grow on trees. However, this theory is disputed by some moles so a second mole-investigator is sent up. But by then it is evening and on the tree there are some cats mewing. Mewing cats, the second mole announces, grow on trees. Finally, a third 'elderly neurotic' mole goes to see for himself, but by then it is night:

> Both schools are mistaken, the venerable mole declared.
> Birds and cats are optical illusions produced
> by the refraction of light. In fact, things above
>
> Were the same as below, only the clay was less dense and
> the upper roots of the trees were whispering something,
> but only a little.
>
> And that was that.

edu/~paw/archive_new/PAW03–04/01–0910/features3.html, accessed 3 August 2006. **31** Kurt Vonnegut, *A Man Without a Country: A Memoir of Life in George W. Bush's America*, ed. Daniel Simon (London, 2006) p. 15.

Ever since the moles have remained below ground:
    they do not set up commissions
    or presuppose the existence of cats.

Or if so only a little.[32]

Most conveniently, the poem's last line allows me to 'wander off', to escape from this essay, taking refuge in a final, 'tartaglian' cipher: science and poetry are not, after all, so different; or, if so, only a little!

32 Holub, *Poems Before and After*, p. 144.

The Wandering Wood

RANDOLPH HEALY

> 'Oh strong-ridged and deeply hollowed / Nose of mine! what will you not be smelling?'[1]

A poem can be anything. As I write this our three year old daughter, Beatrice, is coming back from our hen-house with a single egg in her hand. Miraculously, it arrives unbroken, a feat which robots find surprisingly difficult to emulate. Never mind the uneven terrain, egg after egg is dropped or crushed as they under- or over-estimate the degree of strength required to hold them. Beatrice's solution is to keep the egg on the point of falling, continually releasing her grip until micro-slips between the shell and the whorls of her fingertips cue her to tighten her hold infinitesimally; the process, as is typical of living things, is permeable to information flowing in both directions.

> 'A fool sees not the same tree that the wise man sees.'[2]

While she would more likely prefer the rôle of the tree to either the other two parts – 'wise man' or 'fool' – I can hardly guess what Bea sees as she picks her way past the hawthorns, down the slope back to our house. For a long time I found the fact of empathy, or even basic conversation difficult to account for. I had been working for some years with a pocket calculator (computers were way out of my financial reach at that stage) trying, on the one hand, to bring my intuitive and intellectual accounts of the world closer together, and on the other to put myself in the position of having new (to me) intuitions. 'There are more stars in the universe / than all the words ever spoken, / as many stars as atoms in a matchstick.'[3]

An earlier version of this essay was published in *Poetry Ireland Review*, 73 (Summer 2002), 115–29.

1 William Carlos Williams, 'Smell!' from *Al Que Quiere!* (1917), in A. Walton Litz and Christopher McGowan (eds), *Collected poems, vol. I, 1909–1939* (Manchester, 1987), p. 92. 2 William Blake, *The Marriage of Heaven and Hell*, in Duncan Wu (ed.), *Romanticism: an anthology* (Oxford, 1994), p. 87. 3 Randolph Healy, 'This size of this universe', in Richard Caddel and Peter Quartermain (eds), *Other: British and Irish poetry since 1970* (Hanover, NH, 1999), p. 114.

Yet, though the edges of our universe are expanding at almost the speed of light, the place is so big that, proportionally, Beatrice is growing billions of times faster. With this degree of elbow room even a planet can seem quite inconsequential. If the entire Earth disappeared in a puff of smoke, the loss of matter would be no more noticeable than if an atom fell off your body. What of the human dimension? Taking all the people who ever were, in all of history, estimated to be about a hundred billion people, the crowd wouldn't fill Leinster, a province of Ireland, itself a crumb off the Western coast of Europe. Nevertheless, I find Russell's description of himself as a 'tiny lump of impure carbon and water impotently crawling on a small and unimportant planet'[4] more convincing as theatre than as philosophy. There must be some way of looking at this child so that her extraordinary nature is not erased by mere scale.

'How often / in this loneliness, / unlighted'[5]

My mother was born in Belfast in 1920. Her family had to move to Scotland to avoid the violence of the time, her parents' mixed marriage making them a target on both sides. My father, born three years later in a tenement in the capital of the Irish Free State, saw several of his siblings die. After his mother had a breakdown her children were taken into care. Conditions were scarcely less harsh in the orphanage to which my father was assigned – inexplicably, and heartlessly, his brothers and sister were sent to different institutions. His unremitting 'that was very tasty' after every dinner, no matter what was on the menu, used to puzzle me until I found out that he used to have to run his meal, stale bread, under a tap in the orphanage until it became possible to chew it. As soon as he was legally allowed, he left school and went to work as a messenger boy for the post office. My mother's family had settled in Scotland and were very happy there. However, with the commencement of the Second World War, she was given an ultimatum: either work in the munitions factory in Ardeer or be deported back to Northern Ireland. Not wishing to lose her family, friends and work, she stayed. It must have been an extraordinary place, a great park full of small huts about a hundred yards apart. Every now and then a hut would go up and she would have lost another friend. I don't know how she stuck it. My mother and father were originally pen friends. When they married, the question as to which country they would live in arose. It was not settled immediately, and for years they lived apart. In the end, after my elder sister and I were born, she reluctantly moved to Ireland. I don't think she was ever really happy there, and whenever she said 'home' she meant Scotland.

4 Bertrand Russell, *History of Western Philosophy* (London, 1980), p. 13. 5 Eavan Boland, 'The Woman Changes Her Skin' in *Collected poems* (Manchester, 1995), p. 86.

> 'and ourselves / So infinitesimally scaled // we could fit through the eye of a needle'[6]

I liked school, and English and Mathematics in particular. Our class of 48 pupils in the primary school had, unusually, the same teacher for all six years, a kindly, well-read man, Jim O'Sullivan, from Cork. Though even he could not entirely prevail over the insular tendencies of the Irish education system. For example, I remember in 1966 we had been studying the 1916 Easter Rising for weeks. 'Gold banners between houses in every town.'[7] (For the mathematically catatonic I will point out that it was the fiftieth anniversary of the rising and it was celebrated in the most uncritically adoring terms everywhere. I felt a little left out as my mother ignored it completely as something merely and contemptibly Irish, and my father hardly said a word anyway. My friends all seemed to be learning the correct feelings to have about this great event at home, while I was left all at sea.) Eventually our puzzlement swelled to the point where somebody asked the teacher why the leaders were so summarily shot. It seemed too harsh. Mr. O'Sullivan got a distant look in his eye and said, 'Well, you see, there was a war in Europe at the time.' This was the first hint we'd heard of World War I. And we only knew about World War II because of films and the comics: 'Die *Britisher! Achtung! Schnell! Schnell!'* And so on.

> 'my telescope leans blind against the wall / its mirror cataracted with fresh dust'[8]

About this time it was discovered that I needed glasses. But when I got them, I couldn't see a thing, so I developed the habit of looking through the gap between my nose and their metal bridge. This required some concentration and, I suspect, quite a degree of facial distortion, as my teacher and my mother asked me what was going on. I explained my difficulty. My mother, who could be quite a sport, took me at my word and made another appointment with the ophthalmologist. Said august personage refused to countenance any possibility of error and ticked me off for not wanting to wear them. My mother then brought me to a private optician who discovered a plus where a minus should have been and explained that the glasses were four times too strong for me. They must have been hummers because what I normally wear are real bottle ends. The new pair worked like a charm, and for the first time my world came into focus. I wasn't entirely sure I liked this, but at least it took the guesswork out of getting a bus.

6 Seamus Heaney, 'The Railway Children' in *New selected poems: 1996–1987* (London, 1988), p. 159. **7** Dermot Bolger, '1966' in *Taking My Letters Back* (Dublin, 1998), p. 51. **8** Trevor Joyce, ''93/94' in *Stone Floods* (Dublin, 1995), p. 37.

I began secondary school and then things went a bit pear-shaped. I became ill and was bedridden a good deal of the time. My medication had side-effects which included disrupting concentration and causing memory loss to various degrees. This made reading next to impossible as I found it extremely difficult to remember what it was I had just read. In the end I left school at the age of fourteen. Having no qualifications the range of careers open to me was limited. My first job was as a sales assistant in a jewellery shop. The shop employed two watch repairers and I found their work fascinating. By and large the days were pleasantly uneventful. Though one time I barged into the back office without knocking and disturbed the boss who was busy cracking teeth for their gold fillings. Some of the teeth appeared to have some kind of, well, flesh attached to them. Another day we heard a loud bang. My immediate reaction was that it was some kind of bomb. This turned out to be correct. The shop was closed and we started for home. There were no buses and I had to make my way on foot. I remember running down interminable faded Georgian terraces, unable to take my eye off the buildings, as if mere looking would keep them up. After a year I left this job with the idea of taking correspondence courses, but the difficulties mentioned above soon returned me to the workforce where I successively took up the roles of Hoffmann presser, telex typist, security man and housepainter. 'I'd make a first rate detective. / Except for the bits that involve courage. / and expertise. and enthusiasm for the whole thing'.[9] After four years I was able to return to school and then attended Trinity College Dublin where I studied Mathematical Sciences.

'Logic is a mechanism / made of infinitely hard material'[10]

I'd been writing poetry from the age of twelve, mainly short intense lyrics. My final year in college was a watershed in which my conception of poetry expanded considerably. The catalyst for this was the experience of hearing Maurice Scully reading ten or eleven times as part of a road-show organized by Dermot Bolger. I liked Maurice's work immediately, especially its flexible line. He suggested to me that I consider writing a poetry of ideas, which led me to consider strategies derived from mathematical logic. In his *History of Western Philosophy*, Bertrand Russell time and again criticizes philosophers for not following an argument wherever it may lead. The scholastic philosophers are particularly scolded for their habit of bending the rules in order to 'prove' statements which their beliefs require them to hold true. He was, of course, being more than a little unfair. Still, thanks to them logic is often associated with unyielding dogmatism, duplicity even. Euclid's twelve-volume *Elements of Geometry* was their model. They believed that his geometry was a faithful transcription of reality, and that its form – using a very

9 Peter Riley, *Untitled Sequence* (Bray, 1999), p. 5. **10** Steve McCaffery, *Evoba* (Toronto, 1987), p. 14.

small number of 'self-evident' propositions and the rules of Aristotelian logic to prove, or even generate a vast number of geometrical principles – would transfer to theology the irrefutability of mathematics. There was also a belief that geometry was not empirical, that its results could be deduced entirely from reason. This could be taken as a metaphor for revelation, something independent of experience. Pythagoras had attempted a similar cross-breeding of religion and mathematics. Overwhelmed by his discovery that musical intervals could be described in terms of ratios of natural numbers, he decreed that all of nature was rational and that all its properties could be so expressed. Very soon it was discovered that the square root of two cannot be equal to any ratio of whole numbers.[11]

Knowledge of this was suppressed, and violence, even murder, done to those who tried to make it public. But they couldn't keep a lid on it. Nor could the scholastic theologians prevent the discovery of non-Euclidean geometry and non-Aristotelian logic. I had no interest in logic as a machine for cranking out absolute truths or for riveting one's beliefs together. Indeed, in their failed attempt to deduce mathematics from purely logical considerations, A.N. Whitehead and Russell showed that logic was a far more interesting entity. Paradoxes emerged which seemed to be woven into the very fabric of the discipline.[12] Logical systems were later shown to be incomplete, that is, not every proposition can be proved to be either true or false. They were shown to be incapable of proving their own consistency without the aid of a meta-system, which would in turn require its own meta-system, and so on. A key idea in analytic philosophy used to be that the fundamental principles underlying reality were simple, self-evident and hence irrefutable. A sure recipe for dogma. The exotic paradoxes laid bare by modern logic, inescapable, impenetrable, and insoluble, liberate the discipline from the chain theory of truth. There is no starting point. You may start where you will.

Apart from its epistemological instability, I was interested in the formal and contentual possibilities that logical form opened up. Following an argument wherever it leads can bring you to unexpected places. The sentence structure itself can allow one to be open to ideas which ordinary syntax would fail to suggest. Then again, its organizing principles allow for a music of ideas, where music is

11 The proof of this is beautiful. Suppose $p^2 = 2$ and $p = a/b$, where nothing greater than 1 divides both a and b evenly, i.e. the ratio has been 'broken down'. Thus $p^2 = a^2/b^2 = 2$, that is, $a^2 = 2b^2$, which means that a^2 is even. Thus a is even (since the square of an odd number is odd), hence there must be a whole number k so that $a = 2k$. Squaring we get $a^2 = 4k^2$. From the penultimate equation we have $4k^2 = 2b^2$, which can be simplified to yield $2k^2 = b^2$. Thus b^2 and hence b is even. This means that both a and b can be divided evenly by 2, which contradicts the statement that the highest number that does so is 1. Thus the square root of 2 is irrational. This proof was known to the Greeks. **12** The most famous of these concerns the set of all sets which do not contain themselves. Russell gives the example of the set of all teaspoons. This does not contain itself, since it is not a teaspoon. Put all such sets together into one enormous collection: the set of all sets which do not contain themselves. Does this giant contain itself? If it doesn't, well it should, since its job is to contain all those sets which don't. But if it does then it shouldn't because then it wouldn't be a set that didn't contain itself. They're still at the drawing board on this one.

not just noise (which I'm all for) but the arrangement of relationships in time. It also fostered clarity and density. If you think of me bumbling along, periodically half blind and forgetting everything, the attraction these held will be obvious.

Within several months of all this, Maurice Scully published my first book, a chapbook called *25 Poems*. Its contents were divided between the logical pieces and the more obviously traditional memory poems. Both types have plenty of precedents in the history of poetry, Lucretius offering an obvious example. I wished to move on and the next step was to begin working on calculator poems. Aesthetically, this seemed a straightforward development, going from logic to arithmetic to science. Personally, I wished to find alternative vantage points to that offered by my frequently overwhelming personal history. The spirit behind the project was one of expansion, not erasure.

> they bounce back from the screen
> clean and ready
> making parts of the body
> an aesthetic obligation
> before the skeleton bursts out[13]

The first entry in the large spiral bound notebook I kept for the next eight years concerned the masses, diameters and thicknesses of the various Irish coins. This was followed by a long series on cells, starting with a quote from R.D. Laing: 'I, as I write this [...] am a collection of cells (in the order of 2^{64+}) all of which count one cell, forty eight years ago as our common ancestor.'[14] It quickly became obvious that this many cells would fill a cube around one hundred feet in diameter. Thus I revised the estimate downwards to 6×10^{12} cells, welcoming the presence of error and uncertainty from the start. I went on to consider astronomical and geophysical questions. Here is an example of the kind of poem that resulted from these activities:

> *World War II*
>
> Fifty five million people were killed
> at a cost per corpse of over
> a quarter of a million dollars,
> a third their weight in gold.
>
> Which took a total firepower of three megatons.
> Which is the energy

13 Tom Raworth, 'Emptily', in *Clean & Well Lit* (New York, 1996), p. 63. 14 R.D. Laing, *The facts of life* (London, 1976), p. 26. This is a beautiful book and I'd not wish to give the impression that the only thing I got out of it was a mistake.

of a seven-minute hurricane
or of one hour of the world's tides.
Who got a mention?
The history book names 117
or one in half a million.

The advantages
  a rise in technology
    massive development in agriculture.

Indeed, by the equivalence of matter and energy
the firepower condenses to the mass
of a small potato.

How much does this represent
of all the energy used in human history?

Compare an electric fire to an earthquake
or a full stop to a small dog.

The graveyard would cover an entire city.
The gold would fill

In the course of this I needed to find the price of gold in 1945. It seems that its value was set by the President of the United States, who arrived at the figure by asking his wife to guess a number before they went to sleep at night.

'Perhaps she dreams Freud loses all his money / in a telepathy scam'[15]

After working on purely physical features for so long, I became interested in the size of the mind. Estimates of the flux of information through the senses are huge. Whether one considers the number of rods and cones in the retina and their powers of resolution, or the cellular economy of the eye, which is sensitive enough to detect a single photon in complete darkness, or the number of sensory neural connections and the rate at which they fire, the amount of information entering the eye appears to be of the order of one gigabyte per second. (Including the other senses using similar methods does not increase this estimate appreciably.) I find this very exciting, especially in a society where we are more concerned to evaluate than to value people. What a piece of work is anyone.

15 Penelope Shuttle, 'Tigers', in *Selected poems, 1980–1996* (Oxford and New York, 1998), p. 94.

'The point of music is constantly vanishing'[16]

The obvious question was, where does it all go? How much do we remember, at any level? I read quite a number of books on memory in pursuit of this question. Unfortunately most of them seemed more concerned with forgetting. I was intrigued by Wilder Penfield's findings. The brain, being devoid of pain sensors, can be operated on directly, without any anaesthetic. One account explains how:

> While operating on epileptic patients, Penfield applied electric currents to the brain's surface in order to find problem areas. Since the patients were awake during the operations, they could tell Penfield what they were experiencing. Probing some areas would trigger whole memory sequences. For one patient, Penfield triggered a familiar song that sounded so clear, the patient thought it was being played in the operating room.[17]

In other words, patients didn't just 'remember' experiences but re-lived them, re-experiencing all the sights, sounds, smells and other sensations as well as thoughts and emotions. I remember reading too of a woman admitted to casualty, apparently delirious, speaking in an unknown language. By chance someone recognized that she was speaking Ancient Greek. It turned out she worked as a cleaner, one of her clients being a Classics student who used to recite Homer's *Iliad* and *Odyssey* in preparation for tests. Finally, the state-of-the-art levels of virtual reality that the average person routinely achieves in dreams, far outstripping anything silicon has done, suggests a higher capacity for memory than one might ordinarily imagine.

The brain, representing only 2 per cent of the body's mass, uses 20 per cent of our glucose intake. Only 10 per cent of this is required for cellular subsistence. Suppose the rest is devoted to cognitive activity. Taking the energy required to process 1 bit of information as 10^{-19} joules (roughly the energy of a visible photon, or the energy transferred by one molecule of ATP, or the minimum energy required to cross a synapse) this would lead to a mental activity level of 10^{20} bits per second. Various different starting assumptions lead to a similar value. This is considerably bigger than the sensory flux. It is as if each second the brain generates the equivalent of decades of experience.

Steven Rose estimates the activity of the conscious mind as 100 bits per second. The comparison of this to the sensory flux and mental processing is even more striking, something along the lines of thimble, cathedral and ocean. Which brings me back to the question raised earlier: how is communication possible?

16 Rosmarie Waldrop, 'Conversation 20: on pattern', in *Reluctant Gravities* (New York, 1999), p. 79. **17** http://www.pbs.org/wgbh/aso/tryit/brain/cortexhistory2.html, accessed 3 August 2006.

Given the amount of processing which so vastly outstrips the senses, what is there to stop us disappearing into our own private ocean? Indeed, what is there to stop us abandoning any semblance of reality for a private fantasy? Clearly this is an option. But what prevents it from being obligatory?

One answer is to consider that the mind is itself a natural phenomenon. Our division between the artificial and the natural, while having profound (or profoundly shallow) political implications, may be something of a fantasy itself. Perhaps a poem, sexual preference or a dinner menu can be no more artificial than a spider's web. Thus our worlds, each a work of art, must at least begin in the real. This is unsatisfactory in many ways, and does little to answer the sceptical philosophers.

Another, less trivial, answer is to consider the human population. There was a time, not so long ago, when humanity had the status of an endangered species. (I'm not suggesting that this is no longer true.) The world population would have been numbered in the thousands. Since the family tree of any individual expands as you go back in time, yet the overall population contracts, this means that branches must be shared. We are far more related than we think. It has been calculated that an arbitrarily chosen couple must be at least fiftieth cousins. I would entertain hopes of bringing that limit in closer. In any case, this would suggest that, for all their explosive power, the way minds construct their versions of reality must have common ground.

> 'terrible fashions / consciousness / fashions consciousness'[18]

Given the grandeur of the human mind, 10^{20} bps in the model above, one might ask why consciousness is so relatively puny. Why a bubble floating in all that cognitive space? Perhaps we need to reduce our interior awareness, to avoid complete paralysis in the face of the simplest decisions. A highly filtered subset of the sense flux is prioritized, disproportionately in terms of magnitude, but not in terms of survival. In positive terms it is a lifeline, ensuring we have some connection with 'reality'. Obviously, it is important not to be lost in one's own innards if there are tigers on the prowl. But, as with pain, sometimes nature overdoes things. Our insensitivity to our own vastness causes us to persistently underestimate people, especially other people.

> 'I say these things not because they happen / but because many things happen'[19]

18 Catherine Walsh, *idir eatortha and making tents* (London, 1996), p. 32. **19** Lee Ann Brown, *Crush* (New York, 1993), p. 13.

One of the things I like about the calculations above is the way they treat error. The assumptions behind them are an interesting mix of the precise and the speculative. For instance, one of the variables in standard calculations of the probability of the existence of intelligent life on other planets is particularly fuzzy. This can lead to the conclusion that the Earth is an isolated example, as if life were so improbable that an entire universe is required in order to have one sentient planet, or that we inhabit a packed cosmos in which one can hardly get a word in edgeways. Nevertheless, even at their most speculative, these false facts, number pictures aim for clarity, even if that means being clearly wrong.[20] And this clarity is a rich source of further questions. Small errors in the transcription of genes fuel evolution. Without error the process would grind to a halt.[21] Without uncertainty, matter would collapse.[22] It's a small step to consider randomness as the engine of the world. However, diversity is expensive. Nature is far from green, is indifferent to waste and, like doctors, it buries its mistakes. One reason living beings appear to be so perfect, so exquisitely designed, is that the construction lines have been erased. As Jacques Monod puts it: '[a] *totally* blind process can by definition lead to anything; it can even lead to vision itself.'[23] Yes, but at the cost of deaths on a massive scale. Evolution is not progress.[24] Without the 'mistakes' that died along the way, there would be no 'successes'. In any case, such parenthetical terms are worse than meaningless and only the present mania for continually comparing the incommensurable gives them any plausibility. Every living thing is priceless, incomparably.

'Million-fuelèd, | nature's bonfire burns on'[25]

20 From time to time I check my own calculations. On one of these forays I discovered that a metaphor used in 'World War II' above is incorrect. I left it that way. **21** Apparently able to decline the offer of evolving, snails will only mate with other snails above a certain temperature. The entire population of snails north of a line through Scotland is thus composed of billions of clones. **22** Negatively charged electrons can be thought of as orbiting a positively charged nucleus. So why don't they just stick together? The diameter of a nucleus is of the order of 10^{-15} m. According to Heisenberg's principle the uncertainties in the electron's position and momentum must have a minimum product. In order to be located in so precise an area as a nucleus, an electron would thus have to have a speed in excess of that of light *in vacuo*, which would violate the theory of relativity. Uncertainty stops atoms from collapsing. **23** Jacques Monod, *Chance and necessity* (London, 1997), p. 98. The quotation is contained in the following fascinating but almost self-contradictory passage: 'With the globular protein we already have, at the molecular level, a veritable machine – a machine in its functional properties, but not, we now see, in its fundamental structure, where nothing but the play of blind combinations can be discerned. Randomness caught on the wing, preserved, reproduced by the machinery of invariance and thus converted into order, rule necessity. A *totally* blind process can by definition lead to anything; it can even lead to vision itself. In the ontogenesis of a functional protein are reflected the origin and descent of the whole biosphere. And the ultimate source of the project that living beings represent, pursue and accomplish is revealed in this message – in this neat, exact, but essentially indecipherable text that primary structure constitutes. Indecipherable, since before expressing the physiologically necessary function which it performs spontaneously, in its basic make-up it discloses nothing other than the pure randomness of its origin. But such, precisely, is the profounder meaning of this message which comes to us from the most distant reaches of time.' **24** Those who regard humanity as the peak of a hierarchy might consider that the process continues and the 'peak' will eventually be superseded. **25** Gerard Manley

That order can arise from random processes is intriguing. Many definitions of randomness focus on the product. For example, a sequence is called random if it cannot be represented by any description shorter than the sequence itself. By this light the sequence H, T, H, T, H, T, H, T, etc. is *not* random. I am not entirely happy with a definition that excludes origins in this way. Clearly this *could* be the outcome of repeatedly tossing a coin. 'Nymph on a sampan nirvana audits til tigress.'[26] Opposites have the interesting habit of turning into each other. Can we know whether a given pattern is the result of chance, is a distortion imposed by our perception or reflects something in nature? This has consequences for how we read. How may we gauge a writer's intentions? Is the poem bounded by these in any case?[27] How far can we go?[28]

'A common-sense / Thumbs-down on the dodo's mode'[29]

When we discovered that our second daughter, Florence, was deaf, we contacted the deaf community and arranged to be taught sign language at home. Our teacher, Wendy Murray, was superb. Gradually, we realized that sign language is prohibited in Ireland.[30] One shocking detail after another emerged. I remember signing with a deaf man, a big strong intelligent chap, plasterer by trade, and he was reduced to tears by the memory of having his hands tied behind his back in school to stop him signing. He is a deaf son of deaf parents and just couldn't keep his hands quiet, no matter how often warned to 'use his voice.' In recent years he has been involved in the independent Irish Deaf Society which, unlike the older National Association of the Deaf, is run by the Deaf. When Florence was about six years old I began to write a long poem called *Arbor Vitae*, which is based on our experience.[31] I wanted every detail to be as vibrant and alive as possible, reflecting the tremendous vivacity of our deaf friends. This meant that any technique used had to earn its keep. The anagrams of the word 'chaos' in the poem's first section have a number of different purposes. Fundamentally, anagrams recall life's movement across generations, as 'letters' in various genes shift position. Here, they also represented the way an overdose of order descends into chaos. Instructions are given for working out the 120 anagrams, mirroring the way those

Hopkins, 'That Nature is a Heraclitean Fire and of the Comfort of the Resurrection', in W.H. Gardner (ed.), *Poems and prose* (London, 1979), pp 65–6. **26** Karen MacCormack, 'Export Notwithstanding', in *Marine Snow* (Toronto, 1995), p. 49. **27** One might note that in current practice if a sublime construction may be put on a text the credit is as likely to go to the critic as the writer. However, any bigotry imputed is placed firmly on the writer's doorstep. All things being equal, this would give the author a responsibility quotient of 75%. **28** I remember in class once a student, pen poised, asked 'What is the correct interpretation of this poem?' It's hard not to be struck at how neatly this summarizes a wide range of issues. **29** Sylvia Plath, 'You're', in *Ariel* (London, 1968), p. 57. **30** Much of the history behind this prohibition is documented in Sarah E. Burns in her MSc thesis, 'Irish sign language in a minority language framework', Trinity College Dublin, 1995. **31** The full text of *Arbor Vitae* is available at http://www.wildhoneypress.com.

in institutions are constantly given instruction. They are also a metaphor for branching, twigs on the word branch. Then again, they act as a musical device, 'the birds in the tree' as Maurice Scully once wrote in a letter. In each section there are several acrostics, each a gesture of solidarity in that they will be silent in any reading, no matter how good your hearing is. They also allow the inclusion of horizontal / vertical, mirroring the Irish sign for 'yes', and commenting on the religious background via the traditional idea of the Christian cross as a sign of contradiction in which these opposing directions meets. There is also the idea of history as the vertical and the present as the horizontal in this.

> 'Jangling the metal of the strings […]'[32]

I am interested in establishing as many dialogues between form and theme as possible. As with natural form – natural in the traditional as well as the expanded sense above – this may imply a simple plan, the details of which continually open onto other details. Not so much an onion as a bush, with a readily apprehensible outline, its finer divisions going in every direction. I don't expect the reader to sit with a magnifying glass and geometrical instruments plotting every tangent. But they're there. And they happen naturally enough. Writing is not entirely conscious. Very few activities of any interest are. A technique like this is an act of faith in the magnificence of the human brain – anyone's brain. Perception is a creative act, far outpacing the capabilities of camera, microphone, and other mechanical sensors. How can one not be poetic with minds so nimble and vast? Anyone who wants to can realize a valid reading of any text,[33] unless they have been convinced otherwise, or unless they have a specific difficulty with texts in general. No matter how many relationships a text may hold, the number is peanuts compared with the intricacies we negotiate as we go to answer the front door. And as with writing, reading does not have to be entirely conscious. That said, while Florence's hearing has improved as she's gotten older, suggesting her deafness was obstructive rather than neural, she is also mentally handicapped, and has taught us that there's a lot more to life than reading.[34]

I would consider technique to be an unreliable fence if you wish to carve up the landscape of poetry. The alternation of prose and verse is beautifully handled in Boethius' *The consolation of philosophy*. Dense argument flows passionately through Lucretius' *On the nature of the universe*. Random methods, whether in

32 Wallace Stevens, 'The Man with the Blue Guitar', in *Selected poems* (London, 1956), p. 52. 33 All the same, I have often been surprised at how many people persist in reading poetry as if it had no rhythm. 34 The Irish association for people like Flo went through an enormous amount of heart searching to find the current term: learning disabled. Neither she, nor her peers, were in a position to comment. It struck me at the time that such concerns, while motivated by care, can go too far. There's nothing wrong with Flo being the way she is. No euphemism required.

divination poems or as part of biology, have an ancient history. Collage was almost *de rigeur* among Renaissance Latinists. The erasure of self was as near complete as it can be among the *Fiannaíochta* poems of the early Irish middle ages.[35] (Though given the nature of mind, I can hardly conceive of any human gesture that does not contain some element of autobiography.) The surreal has to work hard to beat nursery rhymes, never mind the awesome oddness of the simply literal.

'random intelligence echoes planets and stars'[36]

It's late now. When Louise and I were first married we used to read to each other at night, one of the books being Spenser's *Faerie Queene*. In one episode the Redcrosse Knight is stuck in a wandering wood, which allegedly represents error – 'in the tree's underchamber / the roots enmesh and thicken'.[37] Interestingly, woods *can* wander. Even over decades. Many of the world's processes occur on time-scales which are not easily measured by a human pulse. Constellations, genes, languages all drift beyond the range of our ideas of permanence. At the other extreme are processes so rapid, evanescent, yet fundamental, that we hardly guess their existence. Incomplete, uncertain, tenacious, dense and diverse. With a beat. Beatrice is asleep in an armchair, her eyelids fluttering. I'd like to close with a piece that occurred to me in a dream. It's called 'Mutability Checkers'.

The full deck gusts outside the playground
briefly forming an aerial house of cards.
I see a woodlouse chasing a tiger,
and square pegs in square holes.
An atom is the part of your throat that sticks out.
Every Saturday, I am a bicycle.
Famous Dialogues lie on a table.
Enter Socrates, winged by the medium's
dot to dot. Solvent without solution,
ignoramus champ of all history,
I think of you sending away a would-be
empiricist with a flea in his ear,
then sweeping to the end of the argument –
reality as a series of diagrams.
Secateured titan, I dreamed a random river

35 The *Fiannaíochta* recount the exploits of Fionn MacCumhail (Finn McCoole) and his fellow warriors. Their form is rigorously legislated. **36** Joan Retallack, *Afterimages* (Hanover and London, 1995), p. 9. **37** Maurice Scully, *STEPS* (London, 1998), p. 30.

whose surface's inflexions shimmered
with every possible geometry
where all-envisaging blindness hatched
and crossed as chance, swollen with potential,
surged against the given, sculpting a world
where botched and sublime bloomed without design.[38]

38 This poem was first published in Randolph Healy, *Rana Rana!* (Bray, 1997), p. 5. It is the first poem in *Green 532: selected poems, 1983–2000* (Cambridge, 2002).

Matter 15: Water and air words evaporate

MEREDITH QUARTERMAIN

to moisture and siccity,
the leg-bones of horse, ox, rhinoceros
to parched mummies of importance
leaving the vestige of an ear in the fearless
a stump of a tail in the tail-less
a dangling horn in the shorn.

Yes means suppose a void, a doubt.
Something airless flows in the cauliflower –
it's only a matter of time before caverns of eyes,
the oceanic islands, expose organs,
an organic that makes life rendered
has wings of its own.

As reading matter, man writes – marshes
dry as a stick. Biscuit. Bone. The dew of breath
on a glass roars and is roric – how muggy, how juicy
the light of the spoken. Today, rain matters
in gutters. It came in dollars and the spring came
cheap, falling steadily to lip service.

With purpose the organ renders harmful
its function, generally reproducing organs,
each step of reduction a reason to be possible
at the desert station. Seldom can we understand
what we stand on. Soaked up, irriguous, irate
irreversible – sweating ebb tide *al fresco*.

Never sure what path to take – give a slight kick
to uneven – wring out man's rainless extreme –

dusty blows blow solid clinging to probable straws
and play out the obliteration we stand on.
Yet something's remembered
in a letter in a word, retained, though useless,
a clue seeking conclusion.

Technē

HARRY CLIFTON

All rose, all became, in their own way,
Masters of the globe.
                                        And I watched them,
Human being that I was, outstripping me
By degrees, in field and classroom,
Grasping, so quickly, the calculus of change,
And later on, in great laboratories,
Stationary, all hand and eye
As the catalyst dripped in, and the colour changed
And the white magic of fume cupboards
Zoomed upwards – *poof!* – like a mushroom cloud.

Already, they had it over me.
Method. Accuracy. The suspension of feeling
For the matter in hand. Laughter was for afterwards,
Wisecracks, and the lines from Oscar Wilde
Who had an answer for everything.
The *Eroica* symphony – music while you work –
Played through a screen above us, raising the tone.
There were women – colleagues and assistants –
In starched white coats, already deckled
With the yellow of titrations,
Bright asbestos gloves and plastic goggles
Not concealing, for an instant, their femininity –
But that was for afterwards.
                                        Afterwards has arrived –
They have moved, the masters of institutes,
To the promised lands, and the countries of the future.
Cities are for touchdowns,

Conferences. God observes their Sabbaths.
One of them dropped me a line the other day.
'I once saw Man as organs, now as enzymes …'
Not, mind you, that I envied him,
I, with my jar of pens and my Olivetti,
Dealing, like an alchemist,
In concentrates, precipitates, and the retrograde heaven
Of my very own angel, nailed to the wall.

Slow Drag

ALLEN FISHER

'The Glory of Albion is tarnishd with shame' (Blake)

An accountant presses light
switches the projector
a *Last Judgement*
on screen takes his tongue
out of his cheek, watches
auditorium eyes to avoid envy.

It is too late
to take the image seriously
genetic modification
already scattered
there can be no wash out

When telomerase performs maintenance
of chromosome ends
by reversal transcription,
engineers call this negentropy or
an immortality, repressed in human soma.
This enzyme produced as

a tumour marker
in oncology diagnostics,
becomes activity analysed
and evaluated using
Telomeric repeat amplification protocol,
a technique for sensitive detection.

This Real-time Quantitative Trap gave
opportunity to further evaluate telomerase
activity to recognise false-positive
results direct from reaction kinetics.
They observed two kinds of results,
a rapid increase of signal from the outset

of Polymerase Chain Reactions in both channels
from partial degradation of
ampliflour primers by
inappropriate DNA polymerase
and primer-dimer formation
during later Polymerase cycles.

'You didn't understand what I just said.'
'I'm sorry.' 'That's not what I meant. You couldn't
have understood what I just said.' 'I know – I meant
excuse me, but how can you presume to know what
I understood? How could you?' 'That's what I mean,
you were sorry. How could you not be?'

His photograph burns me always
conditional continues change never
absolute or autonomous,
and stylised in its way of informing in
what it recalls of collects snapshots
and postcards something

out of a newspaper, catches, or seems
relevant to a thought, accumulation and arrangement,
translated into an archive, becomes a work of
grid of images, fugue-like repetition
elevates the subject, abstracts it into calligraphic
sequences a network of readings

rewrites the emphasis on chance and
ephemera of Fluxus, the indexical procedures
of Conceptual Art, the relish for the common
place, or descended the staircase in the nude,
Still Life and studies of Baader Meinhof,
an immutable mountain,

a grid, coolly rationalised
into nothing but itself
an autonomous mark returns
the repressed, representations of,
blurred and indistinct
collective memory,

epistemological inquiry,
pocket of a revolutionary cinema
a lived reality, investigation
procedures of that picturing,
into imaginary
exhibition, traces potential installations.

Internal consistency enveloped the bazaar
a revival of allegory
as explicit challenge
the palimpsest reification
of quantum theory usurps
each reading to and from
each version of what experience.

Interest in the undecidable
part of the overall project
paradigm shift
marked by appropriation
and site specificity
torn between dialectics
and deconstruction.

Observation and inference
trapped in the same box
heralded as exactness
a truth bannered on the lawn
in sight of a mower
sharpening steel.

What he presumes as knowledge
wrapped in persuasion
reams and rot collide
butt shades abrupt in starkness in biomass

Real-time techniques propose
a basic value used for judgements,

limited in a plateau phase,
where a threshold cycle occurs in
exponential phases of amplification,
ignorant of cultural spacetimes,
He watches himself
so that quantification is not affected
by any reaction component.

Genetic fallacy to infer
from a prospect
of the razor and antirealistic
to apply to a hypothesis
directed at a beginning
of the universe
that they hold true of it now.

In an unrestricted domain social
choice must work for every some
preference profiles inconsistencies and
incoherent search for adequate restrictions
identified, demanded, easily violated
actual situations

prone to inconsist worry centrality
distributional issues in welfare-economic
probables, vote based procedure
unsuitable to get aggregative index
the interests of the less assertive
cannot but be short in need information.

Then descended a sterner scour
with constant adventure, crescendos burn
Mufflers grief the razor edge and sour
pledge, born and not made, with scorn
beside which files and piles of
destitute reefs belie teem of trauma
where flights and power stir above.

Pailing out the shimmer
with suds in shunt and skim
she spans forth or
loop phones terrific fees
appeasement and this commitment
against a mimetic.

Schematic psychophysical parallelism
can suppose experience becomes directly
of events in particular pieces of matter
neuron events correlated with
somatic events and
events in the world out there.

Informational demands of famine
analysis give an important place
to income deprivation which have more immediacy
and ready usability than the more subtle – and ultimately
more informed distinctions
based on capability comparisons.

Visible signifiers in displayed tracks of ionised
droplets concealed coincidence circuits
constructed of clouds controlled in chambers
by counters, showcase that the majority of events,
emerging from lead plates, are individual
can be numbered and given names.

Photographs, taken by poets and
peasants, or newspaper photographers
not easily seen through,
puts each artist to shame
attracts me
to be so much at the mercy of a thing

when she paints
never a definition characteristic of vision
she blurs
to make everything equally important equally
unimportant perhaps she also blurs out
an excess

we can't rely on the picture of
reality corrected
in accordance with past experience, not
enough for us, we want to know
whether it can all be different, a picture
in different layers, separated by intervals

A laminar flow of air ejected from
between layers, made visible by
smoke curls
up into vortex rings in
real spacetime liquid flows
initially in parallel, get in the way

of each other, lose individuality
give rise to instability of motion
of liquid with fractal structures
turbulated all a sudden in
spatio-temporal chaos applicable
to media in shear flows downstream.

Moats tough stream amaze this visit
Persist precarious word wit has income wing
which prevents wither and instant glance

Peach hues art and countenance
Dry stare and stream and depth and mirth
Rust on spray flight offers strength

Towards silence in blindfolds
Certainty, determined, and unchanged
Expected, light reflects from you
He smiled and released himself in reservation
Now my restful raggedness
Takes away every sense except the explicit

He moves in a kind of oblivion
to the needs of love, thought
curved in a split direction, they
speak about it, a mispoured
glass, a shift, the sun
hurts his eyes less than the

sound of dated keyboard
played in a kind of oblivion
repetition of letter sounds
or plastic persuasions wipe
a wet drip from his eyes as the
clouds shift, becoming a convenience
as he moves in and out the door.

A fielded capability
The fluctuation – dissipation ratio –
an increasing function of slow
frequency embedded
onto deeper properties
of granular materials and other related

non-equilibrium subjects such as traffic
flow, flocking, evolving networks
and turbulence.
As each listener, reader,
and viewer
produces
an informed consciousness.

The extent or minimal needs
for this understanding,
recognition that
this may never be achieved.
Aesthetic production,
produced by receiver and not
an experience of beauty,

Many layers of tremble
the Photographer squeezes button
to capture in palimpsest
a future inscription
on a path of thought
that assaults freedom.

The technician's delay proposes a
rapid formation unified
global states proposes difference

a pressure in deep layers, cell bodies,
with a continuous sheet of neurofil
over each hemisphere.

The Accountant turns off the projector
before the trumpets sound
turns to the listeners
in compact able with margin trace crime
shun mild surd
a metal smelt
ash drifted from shower glow.

Alienated by my own practice
over and over
an open door
experiments, beginnings
estranged from the poetry of others
apellation as a decisive act

Creeping Bent on the allotment counters
love's snares spread everywhere
beneath geese alarms over scaffolding
as an egg breaks, Blue on blue,
the killing machine names it, then
straight through radar grids to a target

I rake a token misery of the wizziness
ash beaten residence
stake a lack honour
restored brisk radio:

shot fear
and shot terror.

Interior manifestation of
aspects of awareness the tempo-
spatial role become transient
texture arranged around my
position inseparable from
persuasive complexity

without compulsion painless
from inducted delight
coupled to a series of
transforms on edges
of recurrence
imagined within.

The Geneticist falls in a tube
compensates
pyrogenetic illusion
He batches with hindsight posed fresh
scratch and pursuer pivot
the phase thetic directs a prurient grief
fuses dead roots and shaded cover

The discussion evolved between stable memory
and reliable communication, between a single trapped
cadmium ion and a quantum decay channel provided by
a single photon emitted spontaneously from cadmium
atom-photon entanglement, implicit before this
early measurement of Bell's idea of inequality

violations in atomic cascade systems
to fluorescence studies in trapped atomic ions.
The photon's polarization entangled
with particular hyperfine ground states
in the de-excited atom. The entanglement
directly verified through

What they loosened on this impious
lot what angled revenge
dependent on success to render irreversible
super imperialism implicit in ideology

burn each expectation
a civilisation based on high wages
precondition for achieving high
distinction between private and government
policy become fuzzy control
over academic training of central

bankers and diplomats
to remove the dimension of reality

think and bind
sanction by fury
fire torment transform
financial independence
derivative of a political and cultural
autonomy that cannot last

All engagements of consciousness
involve an aesthetic component,
reiterate the Listener,
the Reader, the Viewer
involvement with the artefact.

Use and function
rather than success and value
provide the criteria
for assessment of art,
varieties of what these provide
explain aesthetic intention.

Snow

MAURICE SCULLY

Snow is one of the best insulators in the natural world.
As layers accumulate the snow matures and forms strata.
Warmed by the earth below, the base begins to lose moisture
by flowing off the tiny rays of its crushed crystals to the colder
crystals of the snow above. Gradually an open space is formed
interspersed with crystals of ice larger than those of snow and
different in shape: hollow pyramids that fuse together at their
tips. Down here in this latticework of interlocking ice columns
the air is warm, moist and still. The light that filters through is a
pale bluish white. The only sounds are the scamper of little feet
or the muffled movement of a predator above. Freezing gales
slam into the forest canopy but down here there is food stored
in the forest earth and seldom any need to visit the world above.
Sometimes, a drizzling rain will thaw the upper snow. When it
re-freezes as a tough layer of ice the exchange of gases between
the ground and the air above is blocked. Carbon dioxide begins
to fill the crystal pathways. When this reaches a dangerous level
the small occupants build ventilator shafts (weasels and owls are
alert to their opportunity in a time of scarcity). Or a passing fox
may catch the scent and hear the sounds that signal the presence
of a living animal below. Or a deer may break through accidentally
with its hooves, causing a little local avalanche. Soon though, the
process whereby the lower layer yields up its moisture to the colder
layers above will start again and in time the corridors below will
be safe again and snug.

From *C++*

DYLAN HARRIS

1.

language has
new
there's

concept my
programming
falls

all the story's heard
on word one
sung

understand
semantic all
read

on
one
word

2.

implications
are defined
precise

you absolute
know
the last

3.

seventy-seven words
fifty-nine symbols

every concept
express
eventual

seventy-seven words
fifty-nine symbols

english
three hundred thousand

seventy-seven words
fifty-nine symbols

eleven duple
two archaied
evolution

english
three hundred thousand

every concept
express
eventual

seventy-seven words
fifty-nine symbols

precised

4.

the language
exact expand

unseen gentlemen
suited for black

ministers of semantic economy
monks of lexical light

5.

run ride bicycle i
rush dawdle walkers
air sprint shock by

sunlight green light
summer forest cool surround
bridleway king

ego high
peacocks' eyes

to be roar low
consume loud
stun

buzzed by the guys
fly low
shock jet technology

but
i thought
i's
one

6.

up down coast
this there that port
that there this port

up down coast
that there this port
this there that port

up down coast
my child me took sail
to cross the out there ocean

up down coast
this there that port
that there this port

up down coast
my child me took sail
to cross the out there ocean

up down coast
that there this port
this there that port

son i'll put together
chance the ocean
expedition

up down coast
this there that port
that there this port

but i rode
with my hope
pressed shut

up down coast
that there this port
this there that port

son i put together
chance the ocean
expedition

up down coast
this there that port
that there this port

but i rode
with my hope
pressed shut

up down coast
my child me took sail
to cross the out there ocean

up down coast
that there this port
this there that port

son i've put together
chance the ocean
expedition

but i rode
with my hope
pressed shut

up down coast
this there that port
that there this port

son i went

my spirit
loss

no coast
no port
no ocean

i lost
i made my
took

dorade

KIT FRYATT

easy to make a mistake
o o
winking at the bream

under the glair
i i
hard like ball bearings

swallow
m
with frascati

nifty bit of kit
i i
the fast net

netting droplets
o o
letting them drop

baffled actually
m m
agrees himself to my right

as to how it works

o o

but I'd like to show

a child despite that

i i

the way dad did me locks

window tax and the plough

m m m

but not the nebula at 3000

light years wide the most massive H

I I

region in the locality

nor its irregular or spiral bar

o i

galaxy known as

the Large Magellanic Cloud

L M C

its blue aureate bounded by

dorade

that I found myself

Notes on contributors

HARRY CLIFTON has published five collections of poems with Gallery Press: *The Walls of Carthage* (1977), *Office of the Salt Merchant* (1979), *Comparative Lives* (1982), *The Liberal Cage* (1988), and *Night Train through the Brenner* (1996). *The Desert Route: Selected Poems 1973–1988* was published by Gallery in 1992 and was a Poetry Book Society Recommendation. A chapbook, *God in France*, was published by Metre Editions in 2004. His stories are collected as *Berkeley's Telephone* (Lilliput Press, 2000). He has lived in various parts of the world, and is a member of Aosdána. *Secular Eden: Selected Poems, Paris, 1994–2004* will be published by Wake Forest University Press in 2007.

PHILIP COLEMAN is a Lecturer in the School of English, Trinity College Dublin, where he is Director of the MPhil in Literatures of the Americas. With Philip McGowan he has edited *'After Thirty Falls': new essays on John Berryman* (Rodopi, 2007), and he has published essays on various aspects of American literature in journals such as the *Irish Journal of American Studies*, the *Swansea Review*, *Etudes Irlandaises*, *Revista de Estudios Norteamericanos*, *Metre*, and *Thumbscrew*. He is currently completing a book on John Berryman for UCD Press.

HELEN CONRAD-O'BRIAIN is a Research Associate in the School of English and the Centre for Medieval and Renaissance Studies in Trinity College Dublin. She has published articles on aspects of medieval literature in *Hermathena*, *Long Room*, the *Oxford Dictionary of the Middle Ages*, and elsewhere. She is a life member of the International Society of Anglo-Saxonists. With Anne-Marie D'Arcy and John Scattergood she edited *Text and gloss: studies in insular learning and literature* (Four Courts Press, 1999).

ALLEN FISHER is a poet, painter, publisher, editor and art historian. He has produced over one hundred different chapbooks and books of poetry, graphics and art documentation. He currently edits *Spanner*, lives in Hereford and London, and is Head of Contemporary Arts at Manchester Metropolitan University and Professor of Poetry and Art. He has exhibited paintings in many shows and examples of his work are in the Tate Gallery collection, the King's College archive, and the Living Museum, Iceland. The work *Gravity as a consequence of shape*, a long work published in many smaller books and booklets started in 1982, was completed in 2005 in three volumes (*GRAVITY* [Salt, 2004], *ENTANGLEMENT* [The Gig, 2004], and *LEANS* [forthcoming 2007]).

KIT FRYATT is a lecturer in English literature at the Mater Dei Institute of Education, Dublin City University. She has published poems, short fiction, translations, reviews, and articles on contemporary literature in a wide range of books and journals, including *India and Ireland: colonies, culture and empire* (Irish Academic Press, 2006), *'After Thirty Falls': new essays on John Berryman* (Rodopi, 2007), *Irish University Review*, the *European English Messenger*, *Poetry Ireland Review*, *REA: Religion, Education and the Arts*, *Metre*, *Skald*, and *Cyphers*. She is currently completing a book-length study of allegory in twentieth-century Irish poetry for the Edwin Mellen Press.

DYLAN HARRIS is a poet and Windows/Unix C++ software engineer working in Belgium. His programming and poetry inform each other, and much of his writing, music and photography can be found on his website, www.dylanharris.org. An authority on Citroën design, some of his recent writing is to be found on the *Great Works* website. In 2006 he contributed to the Cambridge Conference on Contemporary Poetry.

RANDOLPH HEALY studied mathematical sciences at Trinity College Dublin. He has published *25 Poems* (Beau Press, 1983), *Envelopes* (Cambridge Poetical Histories, 1996), *Arbor Vitae*, *Flame*, *Scales*, *Rana Rana!* (Wild Honey Press), and *Green 532: Selected Poems 1983–2000* (Salt, 2002). He lives in Wicklow where he teaches physics to second-level students and runs the Wild Honey Press.

KATE HEBBLETHWAITE is a Research Fellow in the School of English, Trinity College Dublin where she obtained her PhD in 2006. With Elizabeth McCarthy she has edited *Fear: aspects of an emotion* (Four Courts Press, 2007), and she has prepared editions of works by Bram Stoker for Penguin. These include *The Jewel of Seven Stars* (2007) and *Dracula's Guest and other weird stories* (2006).

JOHN HEGARTY was elected 43rd Provost of Trinity College Dublin in 2001. A former Professor of Laser Physics in TCD, he joined the College in 1986 after nine years working in the United States at the University of Wisconsin and Bell Laboratories. He was Dean of Research in TCD from 1995 until 2000. He is a Fellow of Trinity College Dublin, the Royal Irish Academy, and the Institute of Physics.

DARRYL JONES is a Senior Lecturer in the School of English and a Fellow of Trinity College Dublin. He is the author of *Horror: a thematic history in fiction and film* (Arnold, 2002), and *Jane Austen* (Palgrave Macmillan, 2004). With Anne Dolan and Patrick Geoghegan he has edited *Re-interpreting Emmet: essays in the life and legacy of Robert Emmet* (UCD Press, 2006), and with Stephen Matterson he co-authored *Studying poetry* (Arnold, 2000).

BENJAMIN KEATINGE is a graduate of Trinity College Dublin, where he obtained his PhD in 2005. He is the author of a number of articles on Samuel Beckett, and with Aengus Woods he is currently editing a book of essays on Brian Coffey for Irish Academic Press. He is writing a monograph on Beckett and mental illness, and he teaches English literature in the South East European (SEE) University in the Republic of Macedonia.

IGGY MCGOVERN is Associate Professor of Physics at Trinity College Dublin where he specializes in surface and interface physics. He has been Director of Advanced Materials at TCD, Chairman of the Institute of Physics in Ireland, and Chairperson of the Thin Films and Surfaces Group of the Institute of Physics. He has held Fellowships at the Fritz-Haber-Institut der Max-Planck-Gesellschaft, Magdalen College, Oxford, and the Institute of Advanced Studies in La Trobe University, Melbourne. His poetry has been widely published in anthologies and journals in Ireland and abroad, as well as in the popular 'Poetry in Motion' series on the Dublin suburban rail system (DART). His first collection of poems, *The King of Suburbia* (Dedalus, 2005) won the inaugural Glen Dimplex New Writer Award for Poetry in 2006.

STEPHEN MATTERSON has written extensively on American literature. His books include *American literature: the essential glossary* (Arnold, 2003), *Berryman and Lowell: the art of losing* (Macmillan, 1987), an edition of Herman Melville's *The Confidence-Man* (Penguin Classics, 1991), and *The Great Gatsby: the critics debate* (Macmillan, 1990). With Michael Hinds he edited *Rebound: the American poetry book* (Rodopi, 2005). He is a Fellow of Trinity College Dublin, where he is an Associate Professor and Head of the School of English.

PETER MIDDLETON is a Professor of English in the School of Humanities at the University of Southampton. He is the author of *The inward gaze: masculinity and subjectivity in modern culture* (Routledge, 1992), *Literatures of memory: history, time and space in postwar writing*, co-authored with Tim Woods (Manchester University Press, 2000), and *Distant reading: performance, readership and consumption* (Alabama University Press, 2004). He has also written many essays on modern and contemporary poetry. His collection of poems, *Aftermath*, was published by Salt in 2003. Current projects include a book on science and American poetry in the Cold War, a study of British post-war poetics, and a collaborative funded project on the history of poetry performance (see http://poetry.eprints.org/ and www.soton.ac.uk/~bepc).

AMANDA PIESSE is a Fellow and Senior Lecturer in English at Trinity College Dublin, where she works in the areas of renaissance literature and children's literature. Publications relevant to the present work include an edited collection, *Sixteenth-century identities* (Manchester University Press, 2000), and essays on renaissance drama in recent collections including one on *King John* in *The Cambridge companion to Shakespeare's history plays*, ed. Michael Hattaway (Cambridge University Press, 2002) and another on early modern education in *Shakespeare and childhood* ed. Kate Chedgzoy (Cambridge University Press, 2007).

ANDREW J. POWER is a graduate of Trinity College Dublin where he obtained his PhD in 2006 for a dissertation on William Shakespeare. He teaches a variety of courses in the School of English, and he is currently writing a monograph on the neoclassical sources of Shakespeare's tragedies.

MEREDITH QUARTERMAIN is the author of several books and chapbooks, including *Terms of Sale* (Meow, 1996), *Spatial Relations* (Diaeresis, 2001), *Inland Passage* (housepress, 2001),

A Thousand Mornings (Nomados, 2002), *The Eye-Shift of Surface* (greenboathouse books, 2003) and *Vancouver Walking* (NeWest Press, 2005). With Robin Blaser, she recently completed a series of poems, entitled *Wanders* (Nomados, 2002). Her long poem *Matter* is forthcoming from Bookthug in 2008. Her work has also appeared in *Matrix, Canadian Literature, Prism International, ecopoetics, Queen Street Quarterly,* the *Capilano Review, West Coast Line, Raddle Moon, Chain, Sulfur, Jacket,* and other magazines. With Peter Quartermain she founded the small press Nomados.

JOHN SCATTERGOOD is the author of several books and essays on medieval and renaissance literature and he is widely acknowledged as an authority on the poetry of John Skelton, whose poems he edited for Penguin. He is the author of *The lost tradition: essays on Middle English alliterative poetry* (Four Courts Press, 2000), *Manuscripts and ghosts: essays on the transmission of medieval and early renaissance literature* (Four Courts Press, 2006), and *In Leonardo's Garden*, a collection of poems published to coincide with his retirement as Professor of Medieval and Renaissance Literature from Trinity College Dublin in 2006.

MAURICE SCULLY is the author of several books of poetry, including *Love Poems & Others* (Raven Arts Press, 1981), *5 Freedoms of Movement* (Galloping Dog Press, 1987 [2nd ed. Etruscan Books, 2001]), *The Basic Colours* (Pig Press, 1994), *Priority* (Writers' Forum, 1995), *Steps* (Reality Street Editions, 1998), *Livelihood* (Wild Honey Press, 2004), *Tig* (Shearsman, 2006), and *Sonata* (Reality Street Editions, 2006). He has also produced a number of pamphlets, including *Prior* (Staple Diet, 1991; Tel Let, 1992), *Certain Pages* (Form books, 1993), *Over & Through* (Poetical Histories, 1993), and *Prelude, Interlude, Postlude* (all with Wild Honey Press, 1997). With the exception of *Love Poems & Others*, these works constellate *Things That Happen*, a single work composed over 25 years. He is also the author of a bilingual book for children written in collaboration with the German artist Bianca Grunwald-Game, *What Is The Cat Looking At?* (Faber and Faber, 1995). He has also published *Mouthpuller*, a CD, which includes a selection from *Livelihood*, and between 1981 and 1984 he edited *The Beau* magazine and *Beau Press*. He was the recipient of the Macaulay Fellowship in 1981, Bursaries in Literature (1986, 1988), and the Catherine and Patrick Kavanagh Fellowship in 2003. He lives in Dublin.

ROSS SKELTON is a Senior Lecturer in the Department of Philosophy, Trinity College Dublin. Over the years he has mainly worked in the area of overlap between philosophy and psychoanalysis, and with David Berman he inaugurated the MPhil course in psychoanalysis at TCD. He has published several articles on psychoanalysis, most notably in the *International Journal of Psychoanalysis*, and he edited the *Edinburgh International Encyclopedia of Psychoanalysis* (2006). In 2006–7 he read for an MPhil in Creative Writing at Trinity College Dublin, and he is currently working on a novel.

Index